BITTERSWEET SANDS

Twenty-Four Days in Fort McMurray

Rick Ranson

Copyright © Rick Ranson 2014

All rights reserved. The use of any part of this publication — reproduced, transmitted in any form or by any means, electronic, mechanical, recording or otherwise, or stored in a retrieval system — without the prior consent of the publisher is an infringement of the copyright law. In the case of photocopying or other reprographic copying of the material, a licence must be obtained from Access Copyright before proceeding.

Bittersweet Sands is available as an ebook: 978-1-927063-63-7

Library and Archives Canada Cataloguing in Publication
Ranson, Rick, 1949-, author
Bittersweet Sands: Twenty Four Days in Fort McMurray / Rick Ranson.
Issued in print and electronic formats.

Issued in print and electronic formats.
ISBN 978-1-927063-62-0 (pbk.).--ISBN 978-1-927063-63-7 (epub).--
ISBN 978-1-927063-64-4 (mobi)

1. Ranson, Rick, 1949- --Anecdotes. 2. Fort McMurray (Alta.)-- Anecdotes. I. Title.

FC3699.F675R35 2014   971.23'2   C2014-901853-3
                                           C2014-901854-1

Editor for the Board: Don Kerr
Cover and Interior Design: David A. Gee
Author Photo: Fred Elcheshen, Elcheshen's Photography Studios
First Edition: October 2014

 Canada Council   Conseil des Arts   Canadian   Patrimoine
for the Arts   du Canada   Heritage   canadien

NeWest Press acknowledges the support of the Canada Council for the Arts, the Alberta Multimedia Development Fund, and the Edmonton Arts Council for our publishing program. We also acknowledge the financial support of the Government of Canada through the Canada Book Fund for our publishing activities.

NeWest Press

201, 8540–109 Street
Edmonton, Alberta T6G 1E6
780.432.9427
www.newestpress.com

No bison were harmed in the making of this book.
We are committed to protecting the environment and to the responsible use of natural resources. This book was printed on FSC-certified paper.

Printed and bound in Canada

To Isabel Ranson.

We miss you Mom.
Every damned day.

TABLE OF CONTENTS

Shutdown!
Union Hall
Boomers
The Road
Day One – *Orientation*
Day Two – *Selling Time in Ft. McMurray, Set Up*
Day Three – *Starting Over, First Email from Doug*
Day Four – *Secretary Scary*
Day Five – *Toolbox Talk*
Day Six – *Big Mistake, Email Day Six*
Day Seven – *Morning Warning*
Day Eight – *Party*
Day Nine – *Oxygen Content, Emeralds and Snakes*
Day Ten – *Lobotomy's Phone In, That Loud Man*
Day Eleven – *C6, Email Day Eleven*
Day Twelve – *Dinosaur Farts*
Day Thirteen – *Redoing the Job Hazard Card*
Day Fourteen – *Frozen Hams, Email Day Fourteen*
Day Fifteen – *Jason by the Radio, Double Scotch's Issues*
Day Sixteen – *Lobotomy's Final Phone In*
Day Seventeen – *Lunch Break, Email Day Seventeen*
Day Eighteen – *Gas Monitor, a Lesson in Love*
Day Nineteen – *You Can't Drink Him Pretty*
Day Twenty – *Stretches, Terminated*
Day Twenty One – *Too Tall Won't Be With Us Anymore*
Day Twenty Two – *"Got Ten Bucks?"*
Day Twenty Three – *Spider*
Day Twenty Four – *The Great Eastern, Lonesome Road*
Epilogue

SHUTDOWN!

"A bunch of welders went to McMurray for pre-shutdown. They've cleared fifty thousand already. It's going to be a big one."

The voice on the phone didn't wait for, or want, an answer.

"I'll bet they got closer to sixty thousand, maybe even a hundred. Well fuck a wild man, a hundred thou—"

The voice on the phone had to get off. He had to tell more people. As if the telling made him part of the riches already.

I was going to Fort McMurray. From every part of North America: from farms, cities, and small towns with names like Portage La Prairie, Indian Head, Red Deer, and the aptly named Hope, from across the continent, men and women answer that call. They pack bags, fuel cars, kiss their sweethearts with a speed verging on frantic. In their bags they throw enough underwear and socks for a week, shirts for a week, blue jeans? Blue jeans are forever. Everything depends on the workers—their wants, their needs, their safety. Controlling workers is like herding cats. They have to be housed, fed, paid, protected, cajoled, nursed, ordered, and, if need be, fired. Materials can be ordered months, even years in advance; workers can't. Steel can be stored in the snow; men can't. Equipment prices can be discounted if bought in bulk. Men? No, manpower is the great unknown.

These are hard, obscene men who bark when they laugh, because that's what gangs do. Men who are used to working with their bodies, who are not shocked by the sight of their own blood and the blood of others. Men who are accustomed to staying in cheap hotels or construction camps, moving from refinery to refinery, province to province, country to country. Men who drive back and forth down that Trans-Canada Highway, following the work like the Inuit follow caribou.

Before that first worker in a refinery pulls a wrench, shuts a valve, or clicks a computer's mouse to stop the oil flowing, engineers have been planning that twenty-four-day refinery makeover for years.

A scheduled work stoppage is a finely tuned choreography of obtaining materials, leasing cranes and equipment, hiring workers. While the Fort McMurray refinery plans their shutdown, the engineers take into account that they are in competition with all the other shutdowns all over North America, and there's a finite number of supplies and manpower on the continent. What looks to the unschooled eye like rats on a carcass has in fact been planned years in advance. Engineers have made fortunes just developing the computer programs that schedule shutdowns down to the minute.

Orders are sent out to construction companies and unions to supply boilermakers for the pressure vessels, pipefitters for the hundreds of miles of pipes, electricians, scaffolders, carpenters, labourers. And welders, welders for everything.

Planning engineers generate the Work List and the resulting Materials List. The Turnaround Team sends out work packages so that when a man shows up in Fort McMurray, he is assured of a bed and food. Everything comes together in an interweaving of men, materials, tools, camp space, food, and time.

Finally that one day comes when the Turnaround Manager looks around a room at other engineers and all the Turnaround Team members from every section of the refinery. He looks at his watch and orders the closing of a valve, the locking out of a switch, or the turning off of a bank of motors.

The shutdown begins.

The black gold dribbles to a stop, the lights in a thousand sensors dim, the needles in hundreds of gauges freeze. The refinery lies still. All that piping and all those vessels seem to sag, then slip into a lassitude of waiting. The thunderous, vibrating rumble of a working refinery becomes just a memory of an echo. The only sound within the now-sinister labyrinth of pipes and chrome is the faint but constant hiss of heating pipes like a steam locomotive idling in a railway station. Small wisps of escaping steam smell of wet cement, oil, and the rotten-egg whiff of hydrogen sulfide. Men speak in whispers at such times. They look over their shoulders as their steel-toed boots clang in the quiet, echoing on the steel grating. The pipes that once vibrated with the precious liquid now hang limp, glinting dull in the sun like a burnt-out forest of tar-streaked chrome, all right angles and silence.

Before he had clicked off, the voice on the phone had a final excited message:

"After the first shutter, they're going to transfer everybody to the second one, then a third, and on and on. Jeez, man! A thousand guys, seven twelves, maybe fourteens. We'll be buying our own Brink's truck to carry the money. I'm getting on that shutdown. This'll be the biggest shutdown this year."

My truck roared to life. I was hustling out west.

Going to McMurray.

Going to a shutdown.

UNION HALL

I was enveloped by a thing alive. A hundred blue-jeaned men, jostling, clumping, or slouching against white cinder block walls like discarded garden tools in a basement. The throng in the middle of the floor moved and twitched like cattle infested with ticks. The hall shimmered with the electricity of men who were drawn to the protection of the mob, but who at the same time wouldn't hesitate to elbow their neighbour aside for a chance at McMurray's riches.

Men eyed men appraisingly.

I nodded to several men, glanced at a few more, and shot a dark look at a chinless man. The man returned the challenge.

The crowd stood in ragged semicircles facing the dispatcher's desk. To a man, they jammed their hands deep into their pockets, or crossed their arms, hunching, hiding their hands. Always hiding their hands. I pulled my hands out of my pockets, played with them for a while, then shrugged and put them back.

Low murmurs were interspersed with laughter.

"One thing about Daks, you really don't have to ask him what's on his mind." The speaker's heehaw laughter sounded like a dull saw slicing green wood.

I nodded towards the speaker. The young man smiled, his eyes sweeping the room to see if anyone else noticed.

A metallic bong from the microphone echoed through the hall, indicating the dispatcher was ready. Every face turned towards the desk.

His voice surrounded the men.

"Jimco Exchanger, Firebag, night shift, five mechanics, ten B Pressure welders, six-tens, two weeks plus, drug and alcohol test."

A man in front spoke up: "I like six-tens. They give me a chance to get home Sunday, sleep in, wash my clothes, get skinned up."

Men surged towards the desk. One by one, they handed the dispatcher any certificates that could land them the job. The man at the microphone noted each man's information on a white slip. Then he placed the paper into the date/time stamp, and with a loud crack that could be heard in the farthest corners of the hall, the slip was stamped. Another man sent to work. Another contract between union and company made.

Bang! Another man sent to work. Bang! Another, and another, one by one until that group melted away, running home to pack, gas up, and get in a goodbye hump with the wife.

The crowd shrank. The remaining men ebbed towards the desk.

"Golden and Fliese, twenty-four-day shutdown, travel in and out paid, camp job, seven days a week, ten hours a day. Possibility of going to seven-twelves...."

I joined the crowd at the desk.

A man murmured:

"Best-managed company in Canada."

Another jean-clad worker answered:

"Naw, they took those decals off their trucks, too many guys were spitting on them."

A third spoke up:

"It's top-rate and what the hell, it's only twenty-four days."

A fourth spoke:

"Twenty-four very long days."

I leaned towards the dispatcher, who looked so hard his spit bounced.

"Who's the crew?" I asked.

"You accepting?"

I was about to snap a smart-assed reply when I was jostled out of the way by a tall boy, almost a child. The youngster smiled a smile so beautiful I suddenly wanted to hear what he had to say.

"Hi, I was supposed to be here an hour ago, but for the last twenty miles my alternator started to give up. The lights went out an the car started to sputter. So I followed a semi all the way into town. If it wasn't for that semi, I wouldn't have made it."

The dispatcher and I looked at each other, then back to the boy.

"So I gets into the city an' I can't stop, because if I do I'll just stall. So I glided right through the intersections.

"Well, there's this one guy in a station wagon that I must have cut off, and he starts to chase me. I didn't see who the hell he was. All I could see was some guy in a station wagon chasing me.

"So I pull into my driveway with this guy in the station wagon right on my ass. He gets out of his car and starts screaming at me. He was wearing a windbreaker and starts coming towards me, so I pop him one and tell him to calm down. Then he tells me he's a cop.

"So he takes my license and registration and I have to go downtown to the cop shop. I had to wake up my roommate to come and get me.

"So they sat me in this chair and I had to answer all kinds of questions, things that had nothing to do with traffic tickets.

"So I get charged with speeding, dangerous driving, blowing a red light, not stopping for a cop, assaulting a police officer, and operating a dangerous vehicle."

The boy looked back and forth between me and the dispatcher. The dispatcher deadpanned:

"So?"

"So, that's why I'm late."

The young man stared at the dispatcher. The dispatcher's eyes bounced between me and the young man.

"Golden and Fliese?" I asked, returning the dispatcher's attention back to me.

"You accepting?"

I nodded.

The dispatcher gave me a cold smile and, keeping his eyes on me, he addressed the young man. "Go stand over there. What's your name?"

"Doug, ah Doug... Doug Hyland," he stuttered.

"Okay, Mr. Dougdoug. Go stand over there, I'll call you."

My eyes followed the boy. "There goes an accident walking."

The Dispatcher shook his head. "If you can keep him out of jail," he mused.

"Who's the rest of the crew?"

"Well, you got Pops as one of the welders, and for riggers, so far..." He looked at the clipboard. "You got Stash..."

"Shit."

"Actually, he's been pretty good lately. I guess being beaten by half the reserve calmed him down some."

"Well, if you're gonna steal a truck, steal the chief's."

"Then you'll be with whatever comes outta here. Total of twenty."

"Members?"

"Most, let's see... There's Lobotomy, Mongo, Scotch and his old lady Double Scotch, some travel cards, a couple of Permits off the street."

"Piss tested?"

"It's McMurray." The dispatcher gave me a withering look.

"Well, it's only twenty-four days."

Bang! The dispatcher smashed the white slip of paper into the time stamp. Picking up another clipboard, he leaned towards me and smiled.

"Take care of the kid."

"Oh, thanks, thanks a bunch."

BOOMERS

"Are you going to be on my crew at Fort Mac?" the kid exclaimed. There was a long silence as I squinted at the kid through a haze of winter condensation. We stood at the door of the Union Hall.

"I'm going to Fort McMurray," the young man said loudly, as if he were alerting the media.

I watched an old green moving van rumble down an Edmonton side street. I turned and gave the kid a slight nod and offered my hand.

"Rick."

"Doug, Doug Hyland." He grasped my hand. I was surprised at the power of his grip.

"Dougdoug." I smiled.

"So you're off to McMurray." I suppressed a snort.

"Yeah," Dougdoug enthused.

The moving truck was having trouble getting from one gear to another. I watched the truck grinding down the street. When the truck's gears finally thudded into the right set of cogs, I turned back to the boy.

"What's it like?" Being a... a Boomer?"

I studied the kid.

"In there," he said, indicating the union hall, "the guys called you... a Boomer."

This is getting old real quick, I thought.

"I wrote my pre-employment test," the man-child said.

I smiled, a little. "We get so many shoe salesmen telling us how good they are, and when they get to Fort Mac they can't tie a knot. You never know."

"I've never sold shoes."

I chuckled. In the middle of the laugh I caught myself. My face muscles ached from lack of use. "Shoe salesmen is what we call guys that try to fake their way in."

"I'd... like to start booming. Be a Boomer," Dougdoug blurted.

The laughter stopped. I wiped a rough hand over my chin. The gesture was not lost on the boy.

"My Dad does that," he said.

"You're a first-year apprentice. You think it might be a little early to start talking booming?"

"My cousin is a boomer in McMurray and he makes great money."

"I'll bet he drives a huge truck."

"Yeah, it's really nice. It's—"

"And he's got a boat, trailer, and a quad too."

Dougdoug looked at me.

"And he's got alimony and child-support payments."

"Yeah," said the younger man, slow.

"You don't just decide one day to become a Boomer. You just..." I searched for the perfect word, then decided to just blast. "It's not movement that makes you a Boomer, it's skills. You need to be the best welder, the best rigger, the best mechanic on the claim."

"Boomers go from job to job, kinda like cowboys, right?"

I rubbed my face again. "You're not catching what I'm pitching," I said. "Most of them are money-hungry, and they come from everywhere. Listen to the talk in Fort Mac. All they talk about is money. All they talk about. Take a good look at the license plates up in McMurray. It's a real eye-opener."

Our breath hung fat and white in the frozen air.

"Look kid, a Shutdown is like is a football game, a game where everybody gets the shit beat out of them. But nobody lets you stop. There's no time-outs, no rests between quarters. You play offence and defence. The best you can hope for is at the end of the day you aren't hurt. And the noise! The noise is so loud it pounds your ears, face, even your chest vibrates. It's everywhere. It never stops. Stories about how romantic booming is are usually told by people who were never there. And they only tell the good parts."

"How so?" the kid asked.

"You're working on a real bitch of a job. For two months. Then one night you go to the bar and see a naked pole-dancer and an old boilermaker scrapping. Which one are you going to talk about?"

"That really happen?"

"Slapped him so hard they found his false teeth under the band's drums." I smiled at the memory.

"When you tell it like that, Boomers sound like employed vagrants."

"The only thing a Boomer can depend on is his skills and that the job is going to end soon. Boomers gotta keep moving. As soon as guys arrive at the shutdown, they start looking for the next one. I saw a boilermaker that drove seventeen hours through a blinding snowstorm from Thunder Bay to The Pas, Manitoba to get to a shutdown. When he finally got there, he was so tired he staggered."

I shivered, and turned my collar up. The cold and the company was getting to me.

"A couple of shifts into the job, he started to squirm. Something was always wrong. The work was too dangerous, the foreman was an asshole, the hours too short. Come the first layoff, he's gone. He'd heard about a shutdown in Sarnia, and he took off. Left halfway through the shift, didn't even turn in his tools. Down the Trans-Canada, looking for another shutdown. Like a dandelion seed in the wind."

I studied my hand. I'd never worn the wedding ring.

"A Boomer's got no family. He knows a lot of guys by their first names, but nobody they could phone up and get invited over. The closest these guys got to a home is a fifth wheel."

The boy's innocent face was getting to me. I wanted to either shake him or hold his hand.

"You gotta have skills. If you don't have the skills, you can run around all you want, but you are always going to be the last hired and the first laid off. That's not a Boomer, that's a vagrant. And it's lonely, too bloody lonely."

I reached for the door to end the conversation.

"The only thing good about booming is the money."

THE ROAD

The winter sun was hard in the rear view mirror as the truck chased its shadow north. The music played, the tires hummed, trees and farms passed. My world became snow-covered fields sliding behind me like the engine's rumble.

I left the outskirts of Edmonton around noon, heading north. Once I saw that unbroken horizon beyond that crack in the windshield, a shiver of freedom washed over me.

All those telephone lines, white lines on the road, tar streaks on the asphalt, everything disappears into a fuzzy blackness far out in the distance. I imagined getting so small I saw my body going inside that dot deeper and deeper. I saw where the road should end, but I knew I would never reach it. Sometimes the dot disappeared around a corner or down a hill, but it always came back when the highway straightened. It would swing away for a moment, teasing.

Makes you wonder. Was that what it would be like in a spaceship, going into a black hole? Getting darker and darker and smaller and smaller until you were so dark and so small you joined with all the black into a cosmic nothing?

There isn't anything sweeter than the rumble of eight cylinders and four tires on an open road. The sound sank deep inside my heart, like being spellbound by thunderous music in the front row of a ZZ Top concert.

You have to know that at the other end of those chrome tailpipes there's a thousand pounds of metal that means business. When I touched the gas, an angry sound rattled like a Gatling gun.

I played the CDs loud. The songs were raw, dripping of moonshine, smoky bars, and callused vocal cords. I sang along at the top of my lungs to the songs with four beats to the bar. Not much different from beating on a hollow log, but I'm going north on Highway 63, I'm going to McMurray.

Double lanes take the pressure off. Everybody's going the same way, the same speed. It's funny: you inch up on a car ahead, and it seems to go faster. When I passed, I checked the people out: old ladies, businessmen, kids asleep in the back seat, farmers picking farm-dirt from their noses. Once my truck passed them, it seemed to slow down, even though the cruise control had never moved. It's funny.

The double highway ended too soon.

My truck rumbled past a field near Newbrook where workers had nailed hardhats on top of the fenceposts. There's a mile of different coloured hardhats, each put there by people who had no more need of them, and wanted the world to know it.

The oilfields are easy to find—just get gas at Grasslands and turn left at the huge green sign that reads FORT MCMURRAY.

Slowly, that black spot got bigger. It spread along the cracked windshield and the horizon. I turned the headlights on. My world became the windshield, the green and red circles of the dashboard, the music, the heater, the steering wheel.

I passed the place where several years ago I had seen people running beside the highway, the black underside of a car in the tall green grass and a splash of black-red blood on the edge of the asphalt. I remembered the overturned car and the crowds rubbernecking, and a woman in a black blouse and bare white arms holding her mouth in horror at something in the grass.

I see that woman's revulsion-filled face every time I pass that spot.

There are a lot of places like that on Highway 63.

Time drifted like smoke in the wind. The humming of the road became my world. All the miles folded one into another sliding behind my truck, going into that same goodbye.

A car passed doing at least 130 kilometres an hour, then another and another. A four-car convoy of testosterone-dripping mouth-breathers who figure if they travel in packs, the cops can only catch the last one. I moved over and rode the rumble strip and let the knuckledraggers pass.

There'd be nothing wrong with Highway 63 if it only connected two isolated farming towns. If farmers were the only traffic Highway 63 had, it would be a showpiece. But that road doesn't just connect two sleepy towns in northern Alberta. Highway 63, and to a lesser extent Highway 85, funnel all the traffic from an oilfield the size of Florida into a couple of narrow two-lane highways out into to the outside world.

I drove, and drove, through forests of skinny trees with tufts on top of them, like pubic hair on a stick. Woods and vague outlines of lonely farmhouses were the only scenery, black on grey.

Hours I drove. I moved my arms, surprised at their stiffness. I wiggled my toes and looked at my hands on the steering wheel. They were hands that pulled things, held things, and every once in a while punched things. The skin of my hands is covered in welding scars. Once, on a boring afternoon waiting for equipment that never arrived, with the sun hitting my hands and making shadows of the veins, bones, and hair, I had idly tried to count my scars. I gave up when it became obvious that, like craters on the moon, my scars came in so many different sizes that counting them all was impossible.

At first I thought, It's an illusion, but an orange glow defined the night's horizon. I'd seen them before, those domes of orange in the night marking out small towns along the Yellowhead Highway like popcorn on a string.

I had driven through those small towns, catching flashes of people sitting in their living rooms. I wondered what they were thinking. Probably wondering where that loud truck was going. In Grasslands, a couple walked together, holding heavy-gloved hands. I passed back into the night, my eyes adjusting to the lonely black.

But this glow was different. The whole horizon shone like a giant prairie stubble fire, and I was driving into the middle of it.

I shivered. I was doing something, going somewhere. I was entering McMurray proper.

I passed a WELCOME TO sign, and a turnout that I'd never stopped at. There's that hill that always pops my ears, a street sign, a red light. The hospital on the right is massive, its sheer size a warning.

After hours of forest, my truck was suddenly driving through lights and stainless steel highrises.

I stopped at a stoplight, my truck idling, my body shaking from the road. I had driven through the black and come out into brittle lights so vibrant that the buildings shimmered orange.

In Fort McMurray, half-ton trucks drive faster. People walk upright with clear eyes towards some near objective. They are younger, stronger, in a hurry.

Energy squeezes from the cold air blanketing the city. People don't lounge in the oilsands; there's a Monday-morning urgency to their movements.

The work goes on. Winter isn't the pause it is in the south. Snow is bulldozed, huts are erected, creased orange tarps climb to amazing heights to cover the sides of stainless steel towers. Lights in formation pierce the black and orange so the refineries look like Halloween decorations gone mad. Padded workers disappear into their bright orange tents, reappearing only to shuffle penguin-like into other huts.

The work goes on. The still, eternal cold torments Fort McMurray and the refineries, warping sounds, shattering pipes, cracking engines.

The work goes on.

I held the sticky steering wheel and looked over the neon city, an orange nebula in the middle of a silent forest black as tar.

I had arrived.

DAY ONE

(Orientation)

I'll say this about the asshole: he was neat. He stood in front of our class, a turd in an ironed shirt. He looked down his nose at us grubby workers who were obviously beneath his station—a pretty good trick for someone five feet tall with an ass that waddled like a duck.

"You're late! Indoctrination starts at nine AM sharp. Come to the next one." The small man walked towards the criminal coming through the door.

"And the next one starts...?" the offender asked.

"Thursday."

"That's two days."

The intruder had the look of someone who either worked out or worked hard. He was about forty years old, but he didn't have the soft chin like the rest of us. His construction denims were standard but one cut above. The man beside me smiled, nodded towards the criminal, and whispered a soft, singsong warning: "Don't fuck with him."

The instructor's sneer showed that he knew how much a two-day wait in a Fort McMurray hotel would cost. He raised his hand, as if flicking away an imaginary mosquito, and indicated the door. "Be on time next time."

Instructor and intruder stared at each other, each issuing a silent challenge, authority against near-violence. The room went still.

The instructor broke the impasse. "Out."

The intruder looked around the room.

"Everybody who's working for Golden and Fliese, come with me. Out!"

It wasn't a request. The entire room rose. The instructor spun towards the intruder. "You can't leave!" he shouted at the backs of several workers. He turned toward the intruder. "You! Who's your supervisor?"

"He is!" most of the workers yelled, grinning as they stomped out of the room.

Several men thumped down to the reception area. Heads poked out of office doors. The safety coordinator, with blue eyes and a face like a basset hound, emerged from the boardroom carrying a half-finished cup of coffee. "What's up?" he asked, gesturing with his coffee mug toward the departing workers.

"This guy just ordered everybody out," the instructor said.

The coordinator turned to the intruder. "Zat true, Navotnick?"

The instructor started, shocked that the supervisor knew the name of this piece of scruff. "He was late. The rule is—"

"I know the rules," the bemused coordinator said. "I wrote them, Brian."

"Barry," the small man corrected.

The man who had caused the interruption continued. "Without supervision, you'd have a crew of guys getting paid to wander all over the site. Might as well keep them together in a hotel."

"And send us the bill," the coordinator muttered.

"And send you the bill." The man's eyes bore into the small dark man.

The coordinator turned to the men, his face like a mudslide. "Go back and sit down." He spoke like a tired old man shooing his grandchildren off to bed. "Sit down, sit down." To the instructor, choosing his words carefully, he said, "I'm the reason he's late, Brian. He and I had some, ah, points to go over before he could get back on the site."

Barry's neck turned red.

The coordinator grunted. With a tired smile, he continued to wave the remaining class back into the auditorium. He crooked his finger to the instructor. "Acastus? May we have a word?"

"Yes." The instructor almost saluted.

"You could have handled that with a little more... tact, you know, Brian."

"Barry."

"Barry." The coordinator nodded, eyes bluer.

"But he didn't—"

"Ah!"

"—wasn't—"

"Ah!"

"But—"

The older man's eyes became an even deeper blue. He gestured toward the small man with his mug, and a large dollop of coffee splattered onto the white linoleum and Barry Acastus' pants. The man's voice rumbled down the hall and into the auditorium.

"We're not running a gulag, Brian. Barry. Stop being a Nazi."

In their seats, men grinned.

* * *

The day dragged on. The auditorium was hot. The workers had overdressed, the voice of Barry Acastus, the instructor, droned, heads nodded.

"If you cross a red plastic tape barrier, denoting a 'hot' work area, and you don't have permission to be there, you will be terminated. Instantly."

For the twentieth time in an hour, I shifted in my chair. I always attributed my hatred of high school to my hatred of sitting. The only thing that kept me awake through this torture was the pain from my tailbone.

"If you come to work impaired, you will be terminated."

I watched a man across from me yawn, and I yawned back. All over the room, men looked like they were nodding in quiet agreement to imaginary friends.

"There is zero tolerance of discrimination, whether it be sex, race, colour, creed, religion, or sexual preference. If you engage in such activities, you will be terminated."

Acastus stopped. Men on the edge of sleep snapped awake and became wary. Acastus stared at a worker whose head was back, mouth open. He was just beginning a long, slow, nasal snore. The worker next to the sleeping man nudged him.

"Taxi!" the man's neighbour whispered hoarsely.

Taxi noisily wiped his face and stared around, bewildered.

"If you can't stay awake," Acastus snapped, "you'll have to leave. And... you'll have to wait until Thursday to come back."

The crowd awoke. Acastus stared down the now-attentive Taxi.

"Are you going to stay awake?"

The man nodded sullenly.

Acastus looked around the room, like a gladiator about to dispatch a foe demanding the crowd view his power.

"Are you?"

"I'll stay fucken awake," Taxi said.

"That's better."

From the back of the crowd came a stage whisper which both captured the feeling of the attendees, and forever branded Safety Officer Barry Acastus: "This Safety Nazi don't ever learn, do he?"

Only two men in the audience sat watching the instructor. They hung on his every word. Periodically their heads would nod together, as they engaged in a short but animated conversation. My attention shifted from the droning in front of me to the conversation behind.

"That doesn't count."

"Yes it does!"

"It don't."

I turned and looked at the two. An older worker with deep lines in his face and a tuft of pure white hair was talking with a man dressed from head to toe in scruffy black coveralls.

"What's up?" I asked.

"Wait till coffee," said the man in black with a grin.

Twenty minutes later at the coffee machine, the two workers approached me.

"So, Pops," I asked the older worker, "what's the deal?"

The Johnny Cash wannabe interjected. "We've had this asshole before. All this Safety Nazi does is just threaten. That's his only M.O. While he's telling us about the company's harassment policy, he's harassing us."

"I noticed," I said.

"We each put ten dollars in a hat," Pops said, "and the one that gets closest to how many times he threatens to fire us by quitting time wins."

I spoke to the man in black. "What's he up to?"

"Thirty-eight."

"Thirty-eight threats in four hours?" I grinned.

"Shit, I ain't even close!" Stash looked crestfallen.

I smiled. "Let's start a new round. Talk to the other guys. I'm in for ten. Start them off at thirty-nine."

The three of us spread out among the bored workers lounging along the halls and in the smoking rooms of the building. Names, guesses, and a hat were passed, resulting in a sizeable collection. Pops spent the rest of the day looking as if he were carrying a roll of toilet paper in his shirt pocket.

When the session resumed, the class was transformed. Gone were the glassy-eyed stares. Barry Acastus, forever known as the Safety Nazi, was amazed at his sudden command of these men. Never had he taught such an attentive class. He must have thought that he had found the secret: threaten to fire one sleepy worker, and the rest were his.

He laid it on. The more he bore down, the more he commanded us, the more we leaned forward. He ended the class with beads of sweat on his brow, a hoarse voice, and the pride of a job well done. The class almost cheered.

The new apprentice, Dougdoug, won with a guess of sixty-two threats to be fired.

"A personal best," Pops said.

* * *

The classroom portion of the orientation finally came to a close. Now it was time for the tour of the refinery. The entire orientation class clumped onto a line of yellow schoolbuses. The refinery was so large, buses regularly circle the plant to pick up and drop off workers. It takes forty-five minutes for the busses to do a complete lap.

Barry Acastus, The Safety Nazi, stood at the front of the bus, describing the vessels and their functions as they drove past. At one point, he lowered his head and spoke to the Teamster. The bus driver gave him a questioning glance.

"Do it," said Acastus.

The driver left the refinery proper. The structures and the piping became smaller and finally receded into the horizon. The passengers looked at each other and then out at the frozen land. After fifteen minutes bumping down the gravel road, the bus slowed.

"This'll be good," Acastus said to the driver.

The buses creaked to a stop. All around them was a white sand moonscape, sterile, empty as any Arctic winter. As far as the horizon and in every direction, regular mounds of talcum powder white sand gleamed. This was the dumping ground where the used sand from the extraction process was discarded back onto the land. The men who had brought sunglasses hastily donned them.

In the middle of that whitewashed sterility, a three-acre mound shaped like an apple core rose ten feet from the white sand. On top of the miniature mesa, seven hungry trees fought a losing battle to eke out a life in the crackling cold.

"You see that?" Acastus asked softly. "A lot of people think that the company has no regard for the local population. But you see that? That's an Indian burial ground. The company refuses to mine under there. That proves the company is sensitive to the wishes of the locals."

"Jesus," a voice said.

The yellow schoolbus went silent and stayed that way all the way back to the refinery.

DAY TWO

(Selling Time in Fort McMurray)

Fort McMurray is more than giant machines and steam and catacombs of shiny pipes that turn like snakes making love to stairs. Fort McMurray is more than anemic pine trees dying from the roots up, and bewildered moose that regularly stumble out of the woods and right into the middle of refineries. Fort McMurray is more than pick-up trucks with marker flags and dirty chrome convenience stores and parking lots carved out of half-starved forests.

Fort McMurray is that dream of making enough money to become more than what you are. There you can walk taller, talk louder, work harder, get rich. Fort McMurray is Joe Lunchpail's Everest, and ultimately, Fort McMurray is a tragedy.

Fort McMurray suffers from the tragedy of too much. Too many men making too much money too fast. Too many half-civilized men disconnected from their families. Too many men getting treated like criminals by authorities who have good reason to. Fort McMurray has too much, too much of everything.

One of the few things Fort McMurray lacks is objectivity. Bring up Fort McMurray and half the country thinks it's a Canadian treasure, the other half thinks it's a national disgrace. They are both wrong, and they are both right.

The environmentalists sell the past, the oil companies sell the future, and they both slant their message.

Imagine Highway 63 as a tree, with the city of Fort McMurray at ground level. Other than a couple of them, most of the refineries are north of the city. The trunk of that imaginary tree climbs thirty-odd miles high before you pass the first refineries, Suncor and Syncrude. These oldest and biggest refineries sit east and west of Highway 63, like massive gatekeepers on either side of the road north.

Syncrude will always be the poster boy for dirty oil, simply because you can see it from the highway. You come over the hill and there it is, like the Emerald City in *The Wizard of Oz*; Syncrude sits there, in all its chrome and dripping black oil splendour behind two vast settling ponds on either side of Highway 63.

The environmentalists don't even have to get out of their rented Volvos to take an incriminating picture of Syncrude. It's right there.

Chop down every tree, scrape away every plant, get rid of every animal. Dig. Collect fifty feet of oily, dripping sand in one bite, bubble massive amounts of water from the Athabasca River through that black sand, separate the crude, dump the used water in settling ponds so large that they can be seen from space.

The smell of the settling ponds is so pervasive that the construction workers call those squares of froth and stench "Paradise Lakes."

Animals are not wanted. To keep birds away, gay streamers are hung across the ponds, giving the place the look of a used car lot. Air cannons are continuously fired off at irregular intervals. Construction workers go to sleep in the camps to the faint thumping of air cannons in the distance.

Floating on the oily scum of those ponds are forty-five-gallon drums with scarecrows welded onto them, bobbing in the wind. In the summer, the construction workers call the scarecrows "The Newfie Navy." In the winter, frozen in that slushy oil and water mix, they are called "The Newfie Hockey Team."

Dump the sand as white and dead as ground-up skeletons back on some barren spot. Plant trees and moss like Astroturf on top of those huge mountains of inert sand. Spray six inches of topsoil, spread grass seed and mulched hay over the planted trees. Like a scene in some science fiction movie where aliens have created what they think is Earth. Step outside the reclaimed patch, and stretching for miles you see a tortured vista like a World War I battlefield.

This is the prevailing image of Fort McMurray, and it was an accurate image in the 1980s when I first went there. The whole attitude in Fort McMurray in the last century was "I got mine, let God replant." And that is the story the environmentalists sell: that the technology is static, never-changing, locked in the methodology of the 1970s.

And it's wrong. Environmentalists sell the past. When you see pictures in the media of things that aren't there anymore, you start wondering just how noble these environmentalists' intentions really are. Unless a picture of the Fort McMurray is date-stamped and its location identified, I discount it out of hand.

Because the drag-line and truck technology is getting less and less important each year. Think about it. Moving sand is really expensive. Oil companies, being oil companies, only want oil. They don't want the sand, which is a real pain in the ass to move around. And those massive trucks and cranes are not cheap.

After about fifty feet down into the sand, it's too expensive to use the old dig-and-dump technology.

Now, with the new technology, the kind you'll find at the Firebag refinery, all you can see of the oilfields is one-acre squares cut into the forest in the centre of which stands a small wellhead the size of a man. The wells pump steam and solvent into one porous pipe which goes down anywhere from fifty to a hundred feet, then turns horizontally, where a companion pipe sucks out oil loosened by the first pipe. No more dig and dump.

This new technology is pretty much a disaster for the environmentalists, one reason being the local moose love it. I don't know how they did it, but the moose got the oil companies to cut these one-acre pastures all over the place just so they can graze on grasses, plus all these seismic trails through the forest just for them to walk down. And Lord help you if you take a rifle on oil company property.

The moose got it made in Fort Mac. They must know somebody.

Syncrude and Suncor are aware of the optics and are fixing up the tar ponds along the highway in front of their plants. They are landscaping and replanting trees where stinking oil cesspools used to be. It's going to be a real bitch to criticize a place that looks like a park.

To get those shock-inspiring pictures that play to the foreign media so well, they'll actually have to drive their gas-guzzling four-wheel-drive trucks around the back of the refineries.

But if the environmentalists sell a slanted past, the oil companies sell a slanted future.

I can't count the number of times I've seen an oil company executive stand in front of a group of us workers and say: "We have budgeted for..." or "We are in the planning stages of..." or "In the future...."

Well, Mr. Oil Company Executive, in my future, I plan on winning the lottery. I plan on losing twenty pounds. I plan on marrying a rich socialite. Will it happen? Maybe not. But that's my plan.

Call the Oil Company Exec on "The Plan" and it's like you are desecrating holy writ. People spend their careers planning this shit, and demanding actual, concrete startup dates for their "plans" really makes them nervous.

There's just too much money invested in the old strip-mining technology to adjust quickly to the new methods. They just can't do it, but they'll talk about it, and they'll plan it. They'll try to sell the future.

You see, as far as Fort McMurray is concerned, the environmentalists sell a skewed past, and the oil companies sell a wished-for future. If you don't trust either one of them, you just about got it right.

(Set Up)

The skinny woman slammed a sheaf of income tax forms on the desk. Jason looked up from his paperwork. Gwen stood with her white hands on her hips, looking as if she wanted to punch something.

"Twenty men! Twenty men, and every one of them thinks all I want is their little pink bodies!"

"What brought that on?" Jason asked, hoping she wouldn't answer.

"Taxi."

"What did he say?"

"He said that making love to me would be like jumping onto a pile of moose antlers. Only he didn't say 'making love.'"

"Did he say it in front of witnesses?"

Gwen spun around. Barry Acastus stood so close she could smell the ammonia from the starch he sprayed every day on his shirts. She turned back to Jason. Jason puffed out his cheeks and slowly shook his head.

"Well, did he?"

Gwen nodded.

"He's gone," Acastus croaked, eyes glistening.

Acastus hastily started to pull on his parka. He stopped as he was putting on his gloves and looked at Jason accusingly.

"Well?"

"I'm so sorry, Jason," Gwen said.

"Don't be. You don't say that kinda crap and get away with it. Not anymore." Jason looked at Gwen.

"Well?" Acastus repeated.

"Let me finish my paperwork." Jason turned back to his work.

Acastus stood in the middle of the office staring at Jason. "In due time, Mr. Acastus," Jason said threateningly, never raising his voice or his head.

The door slammed as the safety man left. Jason looked at his secretary. "If he's smart," he said, "he'll shut up and let me handle it. If he's as dumb as I think he is, he's running over there right now to make a big speech and fire Taxi in front of everybody."

"Nice way to start a shutdown," Gwen said.

"It is what it is," Jason replied, rubbing his right hand.

There was silence as both people returned to their desks. Jason looked up as Gwen handed him a manila folder with "TAXI" written across the top. To his unspoken question, she pointed to the row of trailers. A lone figure stomped toward them.

"Taxi?"

She nodded.

"Well, we got him fired from his cushy job as an Orientation Training Officer, and this is payback. Any bets that little motherfucker never makes it to the end of the shutdown? Where is he?"

"Probably off somewhere ironing his clothes," she said.

The door slammed as Taxi stomped in.

* * *

Jason appeared in the doorway of the lunch trailer. Conversation ceased as twenty heads turned and watched him.

"For those of you that missed the orientation yesterday," he announced, "my name is Jason Navotnick. On this job, I'll be Supervisor. General foreman duties will be shared between the other two foremen.

"Now, I want to talk about the job, and then there's a couple of things I want to go over before we split into groups. First: as you know by now, the Safety Nazi got reassigned. Word has it that a couple of you fifteen-watt bulbs shot your mouths off about the pool we had going, and as a result, he's lost his cushy job."

"Fuckem," Stash said.

"Yeah, well, like, Taxi just found out, now we got a wounded tiger looking for payback. So watch your ass. Another thing. Giving the secretary a hard time is like making fun of your cook. If I was you, I'd be real nice to the lady that makes up your paycheques. Scary can make your life real difficult in a hundred ways." Jason stared at Stash. "Just to let you know."

Stash put on a show of indifference.

"You don't think so?" Jason countered. "How many unpaid bills you got? Any ex-wives or lawyers chasing you? When they finally track you down, she can get right on that garnishee notice, or she can wait a couple of weeks. Taxi was lucky it was Scary. She spoke up for him. All he got was a layoff. Making a crack about fucking a secretary should get you fired and banned from McMurray for life."

"It was a joke," Stash said, jutting out his chin.

"Yeah, well, find out how she got her nickname. If that doesn't scare you, nothing will."

"What happened?" Dougdoug asked.

"She didn't like somebody's jokes," Jason deadpanned. "All any woman here has to do is say something to the right person and you are gone. And with a charge like sexual harassment, you will stay gone."

Jason flipped the page on his clipboard. Stash glared back at the several sets of eyes quietly staring at him.

Jason put his clipboard down and looked at the crew. "The shutdown, you've already heard, is repairs on the coker. Today, we're on day three of a twenty-four-day shutdown. At midnight on day twenty-four, they'll seal up this coker. Then fifty-five thousand barrels of oil will run through it every day. Anything and anybody still inside this coker at midnight of the twenty-fourth day will end up in the gas tank of some Volvo.

"The coker is the guts of the refinery," he continued. "They'll bullshit you and tell you that they've upgraded their processes, but the fact is, basic refinery technology has never changed. All the refinery does is distil oil, just like they did a hundred years ago. The operators can make adjustments to valves and tanks for a couple of days, but without that coker, this whole refinery will come to a shuddering, constipated stop. So when the company says they'll start up on at midnight on the fifteenth of March, they will start up at midnight on the fifteenth of March."

Jason looked at his clipboard.

"The mechanics, Too-Tall, Shaky, and Roy will be cutting out the places and pieces that need to be repaired. The engineers have marked them with spray paint. However, call before you cut. Talk to your foreman, get his okay.

"You welders? Scotch, Double Scotch, Lobotomy, and Pops. You will be working with, and help setting up, the mechanics. You welders and mechanics, get together."

There was a general scraping of chairs as the welders and the mechanics moved together.

"You riggers?" Several men raised their hands. "You riggers—that includes you, Stash, Mongo—will be hoisting the scrap out of the coker that the mechanics are cutting off, and then you will be lowering it to the ground. You'll stack it on those pallets provided. After all the worn-out metal is removed, you'll be raising the new pieces back in. It's a dirty and dangerous job, so watch your fingers." Jason held up his hand, the one with the missing finger.

"Before we break into teams, everybody helps everyone else. This is too small a shutdown for guys to coast. When you don't have anything to do, the mechanics will help the riggers, the welders will help the mechanics and so on. Everybody pitches in. If anybody wants to play prima donna and stand around saying that helping out is not your job… well then, Slick, I'll make damned sure it won't be. All it takes is a couple of 'Do Not Rehire's to the union and your days of making the big money will stop real quick.

"We got a job to do. We got less than twenty-two days to go. So let's get her done."

DAY THREE

(Starting Over)

Nobody starts a fight in the Tamarack Bar. You are out of your mind if you get into a beef there, and more likely, you'll be out of a job. If you get fired for fighting in the Tamarack, it'll cost you a half a million dollars in lost wages because you'll get banned from every refinery in Fort McMurray. The Tamarack contains some of the hardest characters you'll ever find, but the Tamarack is the safest bar in Fort McMurray.

The oil company, in an uncommon display of common sense, allowed that it was better to allow their workers to enjoy a drink within staggering distance of their construction camp beds than to have a bunch of drunks rocketing up and down Highway 63. The solution was to build a bar adjacent to the construction camp right beside the refinery: the Tamarack.

Originally a double-wide trailer, the Tamarack has a small stage at one end, dartboards at the other, and a standup bar in the middle. But it's not how the Tamarack is set up on the inside that dictates every action that goes on within its walls; it's where the pub itself is located.

Drunk or sober, nobody ever forgets where they are when they're at the Tamarack. Even inside the bar, every once in a while, you can smell that refinery. Everyone knows you can drink at the Tamarack—just don't get slobbering drunk. You can drink at the Tamarack—just don't do it too often. You can drink at the Tamarack—you just better be a happy drunk.

My companion and I had been in the Tamarack for a couple of hours now, sipping beer and staying in that sweet spot between euphoric and drunk. Only the two of us talking, end of the shift quiet.

I'll call him Slim. He was dark and rough-complexioned. His co-workers said he carried moods around with him like an old toolbox. As I remember it, he was a sheet metal worker. We had been on the same job sites, met the same characters, worked for the same companies, but we had never met until now. Beers, loneliness, and meeting someone from home in a strange bar loosened our tongues.

We laughed about the sheet metal worker who we both knew, who once took a dump out on the frozen land without noticing that he'd shit into the hood of his parka—not until he stood and put his hood up.

We talked about jobs where we had both worked. Men had died on those jobs and we talked about the obscene lengths the authorities had gone to protect the companies and blame the victims.

We talked about our families, hopes, dreams. We talked and enjoyed each other like old friends.

The waitress gave us a meaningful glance. Slim and I nodded. She wore stripes. White and black, white and black, white and black, with yellow running shoes. She was short and blonde, chatty and efficient, but nobody's fool. Like the rubber mallets we use to beat metal without leaving dints. With her no-nonsense attitude and dressed in those stripes we instantly called her "The Ref."

Slim and I began talking about divorces. I said that in Fort McMurray, unless you have a couple of divorces and at least one DUI, you haven't worked in Fort McMurray long. I laughed at my small joke but Slim looked away.

Finally he grimaced a smile and started telling a story. He began it in the middle. While he talked and drank, he idly peeled the label from each beer bottle and stacked the labels in front of him. Once he had the labels off, he scraped the glue from the bottle with his thumbnail. Then he would wipe the small wad of glue between his thumb and finger before finally flicking the wad to the rug. Every bottle got the same treatment. Every bottle ended up clean and warm.

I remember that my beer was halfway to my mouth, where it seemed to stay.

"While I was away working up north," he said, "I heard... things. Friends told me that they had driven past my house and I had better get my ass home, if you know what I mean. At first I didn't believe them. Then my brother took me aside and told me the exact same story.

"I didn't tell anybody, but that night, right after work, I hopped in my truck and drove home. I made it look like I was just going fishing. After eleven hours on the road, I drove up and parked close to my house. There was that truck everybody was talking about in my driveway. Just sitting there, like it was making some fucking announcement.

"I walked around the back. Came through the back door and went straight to the bedroom, walked right past them in the bed and got my shotgun outta the closet. By the time the guy was really awake, the barrel of my shotgun chipped his front tooth. He didn't even try. He just lay there with that barrel in his mouth looking like a set of big white eyes along a long, black barrel.

"It was all I could do not pull that trigger."

"I don't know what I'd do," I said.

"I thought about it, and he knew I was thinking about it. I remember the sheets trembling. I let him mull it over for a minute then I said, 'Get out.' He jumps out of bed and tried to get dressed but I jabbed his ass with the barrel when he's bent over, so he got moving, still trying to get his clothes on. I followed his naked ass to the front door, poking him a couple of times.

"I know what he was thinking. He thought I didn't want to mess up my house so I'm going to blow his face all over the front lawn. He started to beg about not going outside. He really started whimpering. To shut him up, I threw his truck keys over his shoulder and they clinked on the sidewalk. He stood there, looking at the keys and back at me, deciding, so I prod him, again. It's January and he's down there naked on his hands and knees on that icy sidewalk gathering his clothes.

"He starts to say something and I raise the shotgun and take a deep breath like I'm getting myself ready to shoot. He started whimpering real loud and backing through the snow with those bare feet, holding his pants in front and the other hand trying to shield his face."

Slim smiled a rare smile.

The Ref came and gave me a questioning look. I nodded, my eyes flickering from Slim for the first time. Slim's eyes never left his memories.

"Pathetic, he looked, really... you know, slimy. The guy skulked, you know? He really looked like a guy who'd fuck somebody else's wife. I couldn't believe, still can't, that this was her idea of trading up."

"What did you do with your old lady?"

"She's got her dressing gown on by now, not that old ratty housecoat, but that new one. That made me mad. I gave her that for Valentine's Day and she's wearing it for him. Anyway she's bent over trying to get her panties up. I grabbed her by the hair and marched her on her toes to the front door. She's making pleading noises.

"She said something about never being home. I was home. Goddamn it! Every three months I was home for more than a week.

"I made sure his truck was gone and then I threw her out. I mean, I really threw her. She cleared the steps, hit the snowbank, and was off running. Her folks lived three blocks away. I went back into the bedroom. The room... stunk."

Slim's eyes stared like he hated the table. His nose flared as he tried to crush the beer bottle with his dirty fingernails.

"I telephoned her parents. I told her father that I just caught their 'pure as the driven snow' daughter fucking my ex-friend.

"He said, 'My daughter is a good girl, she wouldn't do that kinda thing' bullshit. Then I heard their doorbell in the background. He puts the phone down and I hung up. I could just see their faces when she walked in dressed the way she was, smeared lipstick and all."

There was a crash and laughter at the bar. I watched three welders I vaguely knew share a joke. They looked at me and I nodded in return. Slim stared at his beer.

"I had planned out the first part but now I'm shaking. The bed was still warm, for shit's sake. I think right then I had made a mistake and I really should have killed them. Right then I could have killed anybody.

"I ripped off all the sheets and threw them out to the backyard. I got some gas from the snowblower and burned everything. I just kept on going back into the house and getting more and more of her shit, building that fire. That fire was so high it burnt the tree I had planted a couple of years before.

"The neighbours must have wondered. But then again, they must have known what was going on. Nobody called the cops. I guess they wanted to watch. Bastards.

"I stopped throwing things in the fire when I tossed in a box of china. Everything was just a blur. I didn't realize that the china was my Grandma's until it was too late."

"You burned her jewelry?"

"Yeah." He smiled. "She really loved that jewelry.

"By this time, it's morning and I'd been up for twenty-four hours. I phone the Salvation Army and I tell them that if they can be here by noon I'm donating an entire houseful of furniture. They show up mid-morning and within an hour the place was empty. They were happy."

"Didn't you try to sell the stuff? Try to get some money?"

"Too slow. Besides I wasn't thinking about that. I did get a nice tax receipt."

"Nice."

Slim snorted. "So while the Sally Ann is cleaning the house out, I phone the bank and find out how much is still owing on it. Then I phone a real estate guy I know and tell him that the house is for sale for what's owing on the mortgage plus a bit to cover costs. I'd been paying on the thing for years, so it was cheap. The salesman comes over and writes me a cheque. He bought it."

"Isn't there laws about that?"

"Oh, she eventually got half of a discounted house, but she never got the furniture. Besides, the real estate guy made her beg some.

"By this time, it's late afternoon and everything is shimmering because I'm starting a huge migraine. But I got most of my shit in the truck and anything else I couldn't move the agent said he'd send me.

"I'm just about to leave and coming down the road I see my father-in-law. He walked all bent over like he was walking against the wind. When he saw me sitting in my truck, he even walked embarrassed. I wasn't going to make it easy.

"He comes over to the truck and looks at me for a while then says... 'You look like hell.'"

"'What do you want?'

"'She said you jumped to conclusions...'

"'You saw her. They were in bed together, naked. That's my fucking conclusion.'

"He shriveled some more. I didn't think that was possible. He hung onto the door handle with two hands like he was pumping water, looking like he wanted to be anywhere else."

"'She sent me over to get some of her clothes.'

"'In the backyard.'

"'Where you going? West?'

"'Going booming. Could be anywhere.'

"'McMurray?' His eyes weren't looking at me when he said that. You know? When he said 'McMurray,' it was like he saw freedoms just out of reach. He said 'McMurray' a couple of times, like it was the answer to all his prayers."

For the first time in several minutes Slim looked up.

"He thought of McMurray as a chance to start all over, to change his life. You could see it in his eyes. He quit holding onto my door handle and pointed to the house."

"'Door open?'"
"I pointed to the backyard. When he turned, I drove away."

The Ref came back. I shook my head, but Slim nodded. He wasn't finished, but I was.

"Got a young one now. Good girl, young. She does what she's told."

"Still travelling? Gone all the time?" My eyes watched The Ref's receding backside, then I turned and our eyes locked. Slim gave me a long look. He reached over and gathered the beer bottle labels on the table then crushing them into a ball.

"She wouldn't dare."

(First Email from Doug)

From: Doug
To: Dad
Subject: Fort McMurray

Hi from Fort McMurray, where it's so cold when you spit, it crackles. And don't even talk about peeing outside.

The camp I'm staying at has about a dozen computers in the lounge area, so if you wait until supper when everybody else is eating, you can usually get one.

Well, I'm here, working with about a thousand welders, pipefitters, ironworkers, and a dozen other trades on a shutdown. A thousand construction workers, ten thousand tattoos.

Here are some of the things I have picked up in the two days I've been here.

"You know, when I have a cough, I take a laxative."

"Does it work?"
"Yeah, I'm afraid to cough."

The guys I'm working with say that the time to leave Fort McMurray is when you have a favourite restaurant.

One of the crew, a sheet metal worker, told me this: "I was working for a company that maintained swimming pools in Toronto. One day I had a blocked tube I was trying to clear. So I blew into it, nothing. So next I sucked it, and got a mouthful of pure chlorine. Luckily, I didn't swallow any. My mouth blistered up right away. It really hurt. But boy, were my teeth ever white!"

I was talking to Pops, my journeyman. I said to him; "Jeez, Pops! I haven't heard back from the University for two months; my truck is making funny noises; and Stash, the other journeyman we work with, is more than a leeetle strange."
Pops answered, "Never mind. Did you hear about Eric, the other apprentice?"
"No! What happened?"
"He got kicked in the leg by a cow."

I'm having a ball, Dad.

Dougdoug (that's my nickname)

DAY FOUR

(Secretary Scary)

The alarm clock buzzed at exactly 5:30 AM. Gwen Medea slammed the alarm button, knowing the buzzing would wake her neighbours. The walls between the camp dorm rooms were so thin that in the quiet of the night she could hear her neighbour flip the pages of a book. If privacy is what you want, live in a trapper's cabin. Other than that, this is life in a construction camp dorm.

She sat at the edge of her bunk and looked around the room. The room was little more than a cell. It consisted of a bed, a sink stand, a mirror, a closet for her winter clothes, and a small desk with an uncomfortable seat. This glorified closet was her home for as long as she stayed in this construction camp. Gwen Medea could hear distant footsteps—a woman wearing winter boots, she concluded.

Once, someone brought up to the Camp Committee that the size of the rooms in this construction camp did not meet Amnesty International's acceptable standards for jail cells. After a lot of puffing and huffing, the company sent a terse memo saying that the difference between a jail cell and a room in this camp was that here you could leave anytime you wanted. The veiled threat was not lost on the camp committee.

Gwen picked up a towel and, except for rubber flip-flops, she walked naked into the bathroom that she shared with her neighbour.

When she entered, she locked the other door that opened to her neighbour's apartment. Gwen was on her own; however briefly, she was alone.

The only thing Gwen knew about the woman she shared a bathroom with was that she kept it clean, which was the only thing that Gwen cared about.

Camp managers tried to alternate day-shift and night-shift workers sharing bathrooms. As a general rule, they also tried to separate workers by age to prevent clashes. One thing they always did was to separate the sexes. This part of the camp was for women.

Gwen's shower was quick and efficient. It had been a while since she had spent a long languorous afternoon at a spa.

After drying off, she applied baby powder. Powder dries moisture, and perspiration at thirty below is not fun.

She looked at herself in the mirror, pulling and moving portions of her body to get a better look. No spots, no lumps, no discoloration, and the last doctor's embarrassing probing revealed that she was healthy—skinny, but healthy.

"Heigh-ho, heigh-ho," she said to herself, grabbing for the clothes she had neatly laid out the night before. Long wool underwear, wool socks, a T-shirt with a turtleneck collar, blue jeans with the belt her brother the cowboy had given her for her birthday, blue denim shirt, and those $300 steel-toed hiking boots she had splurged on for that walking tour of Switzerland two years ago.

Finally, into the carry bag that always went with her she stuffed shiny black nylon snow pants that zipped up the outsides of her legs. They were a must in case she became stranded and had to walk in the minus-thirty-degree temperature.

Gwen knelt. Putting three fingers on one side of the pure white running shoes, she carefully placed her slippers parallel to the sneakers on the other side of her fingers. She opened her cupboard, staring critically at the stack of identical sweaters, then patted them into a perfect square, making sure the edges of the pile lined up. She set the bottle of mouthwash one finger's width away from the toothpaste tube, which in turn was one finger's width away from the toothbrush. Job finished, she stood at the door and surveyed the room. Sniffing, she refolded and rehung the damp towel. She cocked her head for a moment and concluded that it must be just the play between the lights from outside and the bedroom light's shadow that made the towel look like it was hanging off-centre.

Gwen had once brought a measuring tape to her room to centre the clothes. Halfway through measuring everything in her room, she suddenly stopped. Wiping her hands, she threw the tape measure in the drawer. "I must be nuts," she had muttered to herself.

Gwen had solved the problem by measuring her fingers, and used their width for spacers instead of some cold yellow measuring tape. "It'll have to do," she muttered. "For now."

She put on a down-filled parka that had cost her entire paycheque two years ago. She almost cried when she sewed reflector tape on the parka's back to make it construction-site legal—not fire-retardant legal, but enough for a tiny secretary to pass off as legal.

She shook her head, promising herself for the hundredth time to get rid of this stupid wool cap with the annoying red tassel. Her mother had knitted it for her, and she didn't have the heart to tell her that it, well, sucked. It was too thin, and the tassel hung out the backside of the hardhat, looking like a red wool ponytail.

She put on her leather gloves, knowing that the mittens, each the size of a loaf of bread, which she had sewn to the sleeves of her parkas were always attached, hanging on like she was a three-year-old.

In her pocket was the balaclava that she wore if she ever had to walk the three miles to Golden and Fliese's trailer. She had seen too many frostbitten faces and fingers on people who had put fashion before safety.

She turned and looked at the room one last time. With a sniff, she muttered, "It'll have to do." She repeated her morning mantra as she closed the door and made sure it was fully locked.

It was 6:05 AM when she put on her yellow hard-hat with stickers from a dozen unions and companies and waddled out the door. In the doorway she put on her safety glasses that were mandatory outside every building.

Outside the dormitory, the early morning cold crackled.

Her boots crunched on the hard-packed snow as if they were eating celery. She walked through the black, the vapour from her breath hanging in the air like some long gossamer scarf. Other shapes moved in the dark, but unless they wore very distinctive clothing, she wouldn't recognize them.

The company's one-ton truck stood in a row of identical silent dark vehicles. Gwen liked the way the truck looked. Under the lights of the refinery, the white truck gleamed clean.

She opened the driver's-side door, which groaned with the cold. She slid along the ice-hard driver's seat, her steaming breath hanging in the cab. She felt that this was the coldest time of the day because her pores were still open from the warm shower and she was surrounded by freezing metal. Quickly, she turned the key.

She always felt a small relief when the motor fired.

Gwen waited until the sound of the motor changed from a hard rattle to a rhythmic thump. Every so often, the company would issue a memo restricting excessive idling of motorized equipment and the resulting waste of precious fuel. These memos, usually sent from some sunlit office in a city far to the south, were universally ignored.

She dialed the truck's heater to FLOOR, got out, and, leaving the truck in a nearby parking lot, followed other black shapes to the kitchen.

Gwen glanced at the big wall clock. At 6:15 AM, the camp kitchen was a thousand-person food factory, gushing noise, humidity, and cooking smells. She hung up her overcoat and picked up a damp tray. She joined one of the three lines of workers that inched along, like conveyors pushing their trays along stainless steel rails.

The same creepy kitchen helper smiled at her. She gave him her usual morning grimace. Thankfully, she was kept moving by the twenty people behind her. If some idiot ever stopped the line to ponder the various benefits of bacon over sausages, sure enough, some worker behind would start barking.

Gwen looked for another woman to sit with. No such luck this morning. She sat alone, trying to ignore the conversation of a labourer with a filthy mouth. He wasn't facing her, but the sleaze was meant for her to hear. She burnt her mouth with the coffee in her haste to leave.

Glancing at the large clock above the kitchen doors, she calculated that she had exactly twenty minutes to make the three-mile drive, so she had better make this next unscheduled visit count. She stopped to say a couple of words with an old friend at the security desk, a grizzled ex-RCMP Officer. To an onlooker, they would seem to be two friends exchanging morning pleasantries, unless they saw the old street cop give the filthy-mouthed labourer a hard look over the tiny woman's shoulder.

Gwen gathered her parka and left the steamy dining hall. By now, the idling truck would be warm enough so that her hands would not burn with the cold when she held the steering wheel.

The truck followed a well-worn route. The refinery was constructed in squares or blocks, just like a city. The streets separating the blocks were wide enough for two vehicles to pass, but not much more. She bumped along the laneway so narrow that sometimes passing trucks clicked their side mirrors. Refineries are built to refine and move oil; moving workers from one block to another is a low priority. Trucks don't drive the lanes; they creep.

Gwen stopped behind several yellow schoolbuses unloading men. Masses of dark figures marched into the glow of her truck's headlights and then back into the dark. They walked between the buses and her truck, hundreds of workers surrounding all the vehicles, their breath trailing. The reflective tape on the worker's coveralls flashed like fireflies in a windstorm before they disappeared into the shiny caves of the refinery. She tried to creep forward through the mob. Someone shouted an insult. Gwen almost beeped the horn. Within a minute, the horde melted into the black. The now-empty buses, with Gwen trailing, moved out.

Gwen sighed a small sigh of triumph as she pulled up outside a white trailer right at 6:45. She wrestled with the frozen electrical cord as she plugged the truck into the outlet. The keys to the truck were always left in the ignition in case the foremen needed the vehicle, or it had to be moved in an emergency.

Once, Gwen's company truck was "borrowed," but the truck couldn't go off-site without passing through closely guarded gates. All it took was one announcement on the radio about a stolen vehicle for the truck to be quickly located.

Gwen's boots crunched loudly on the snow's crust. Out of habit, she glanced over the top of the trailers towards the dawn. There wasn't even a hint of the morning sun in the east this early in the year and this far north.

The buildings had been situated there for decades, but the offices still felt temporary. Gwen mused that it was no wonder critics mistrusted oil companies; every office building within the refinery's fence looked like they were only there until something better came up.

She walked into one of the utilitarian white trailers. Sixty-eight feet long, it was a wide-open box except for the Foreman's tiny office. The space was dominated by a large sheet of plywood that served as a desk where the blueprints lay. Several large work schedules were tacked onto the walls. At one end of that white cavern was Gwen's desk. A room divider separated her from the space where the foremen had their desks.

Pouring her second coffee of the day from a pot a labourer had prepared, Gwen turned on her two-way radio, her computer, and a small AM/FM radio that played songs she didn't recognize. Her chair creaked.

Gwen straightened every paper on her desk, placed her two pens parallel, and double-checked the wastepaper basket to see if it was empty.

The clock read 6:55 AM. Gwen Medea, Secretary Scary, was at her desk.

DAY FIVE

(Toolbox Talk)

Hundreds of pieces of reflective tape flickered yellow in the dark, like a line of distant aluminum cans on a conveyor. Slowly, dark figures attached themselves to each glimmering fluorescent badge. Blue-clad men joined in greater and greater numbers, marching purposely, as if on a conveyor belt, along the paths of the stainless steel alleyways. Assembling in groups of ten or twenty, roughly designated by trades or friendship, they all faced one man. Without a command or a shout, the leader began to direct them in morning stretches.

For several minutes they pulled and stretched, bent and twisted. Apart from occasional grunts, they were silent.

After fifteen minutes, with a thumping of dozens of gloves, the exercises were complete. The men broke into trades for the second ritual of the day, the Toolbox Talk.

Several boilermakers gathered in a heated orange tent where men could work on small projects. The enclosure gave the crew a sense of place, safety, community. Hoardings were erected in the first snow of the fall, and remained there until the last melt of the spring. Some hoardings were semi-permanent, year-round shops incorporating shipping containers where tools and toolboxes could be left in relative safety.

The crew shuffled in, waiting for the foreman.

The apprentice spoke up. "Why do we have to do all those exercises?"

"Because they work," replied an old welder.

Several of the other welders nodded.

"Okay, guys," Jason barked, raising his clipboard with today's announcements. "Today's reading is from the Gospel according to Barry Acastus, better known to you scumbag lowlifes as... The Safety Nazi."

"Let's give a clap for the Safety Nazi!" Pops barked. Three or four men thudded their heavy leather gloves together, once.

"Today's Toolbox Talk is about..." Jason paused for effect. "Safety goggles."

The men groaned as Jason read from his clipboard.

"As per standing oilfield policy, absolutely no coloured goggles or tinted safety glasses or lenses can be worn inside any building in the refinery or any other building on this site. At no time will this be allowed to take place."

"Asshole," someone grunted.

"For those of you who choose to not heed this requirement, disciplinary action will take place with the worker being immediately being..." Jason paused. "Shit. He's got two 'being's in here." He showed the closest worker his clipboard.

"Asshole," someone grunted.

"...Being sent off-site and suspended for one shift without pay. Any subsequent violation of this policy will result in termination of employment."

"Asshole," someone grunted.

"Coloured goggles may be worn on the roof area or any outdoor lay-down or fabrication area only. Golden and Fliese's employees and all subcontractors are to diligently follow this directive at all times for the remainder of this project. Field supervision to strictly enforce this policy at all times."

"Asshole," several workers grunted in unison.

"Other than 'asshole,' does anybody have anything to say?" Jason asked, looking up from the clipboard. There was silence in the hoarding. A diesel engine coughed to life close by.

Dougdoug slowly raised his hand.

"Yes?" Jason said.

"I'm kinda new here, but at what point does safety stop and coercion start?"

"Yeah!" several workers spoke up. "That asshole just co-worsted us!"

"He painter-sizes us, like we're dummies!"

Pops was quiet amidst the hubbub, which caught Jason's attention. "Pops? You're the steward." Jason invited the old man to speak.

"You know," the old worker said, "back in the day, getting hurt was just part of the job. We really needed safety back then."

Several of the workers leaned in to hear Pops.

"Now, safety's gotten to be big business. The lawyers all got involved because there's money in it. Be careful what you wish for. Used to be a guy that drank all night and showed up to work the next day was a man. You looked up to a guy that could do that."

"Didn't you have accidents?" the apprentice asked.

Jason held up his four-fingered hand.

"But we were men then," Pops said. "Not like this, writing and signing everything, sneaking around, watching our asses. Little girls firing grown men. You worked hard with men so tough they scared you. Made you proud to be accepted."

"Some of us haven't changed," a welder challenged.

A faraway horn sounded. It was the only sound in the tent.

"Goggles! Let's get back to goggles," Jason said, breaking the silence.

"Goggles fog up," Scotch said.

"Yeah," Pops grumbled, "and you're constantly taking them off with your dirty gloves to clean the fog. On a cold day like today, they're off being cleaned more times than you wear them."

"Do what I did," Stash said, holding his glasses up.

"Jesus Murphy, don't let anybody get a close look at these," Jason said as he handled Stash's goggles.

"What did he do?" Dougdoug asked.

"He's ripped the rubber bottom out," Pops said. "It lets the air in."

"Taking away any safety that goggles are supposed to have," Jason said.

"Well, they don't fog up anymore," Stash grunted, retrieving his goggles. "So what's worse, some air getting in, or stumbling around because you can't see?"

"Don't let Safety catch you," Jason said.

"Don't let Safety catch you doing what?"

Behind the crowd of scruffy tradesmen in the half-open door, Barry Acastus stood. His white helmet glinted in the lights. It looked as if he had pressed his new blue coveralls. His pigskin gloves were so white the pig would have been proud to have donated the leather. His clean coveralls were surrounded by the ripped and oil-stained clothes of the crew.

Noise from the surrounding work area grew in proportion to the silence within the tent.

Barry Acastus, The Safety Nazi, let the orange canvas door swing closed. He was about to say something when the oldest man spoke.

"Acastus," Pops sais, biting off his words, "you can take this memo and shove it up your ass."

"Why?" Acastus looked around in vain for a friendly face.

"Listen, man. We go outside, up on top of the coker off-loading equipment, and you're looking at the crane, right into the sun. We need shades. Then you have to come inside the building to rig that piece of steel. So now, according to this memo, we have to change glasses while we're holding onto that ten-ton load. We're in and out, in and out, all the friggin' time!"

"Yeah," Stash broke in. "You're changing from shades to clear so often with dirty gloves, the goggles get filthy and scratched just changing 'em. Not to mention these things fog up so much they attract dirt."

Acastus looked at Jason for support. The foreman glared at him.

"These things are a piece of shit," Jason spat. "They're okay if you're an office worker out for a stroll. But the moment you start actually doing any work, they fog up."

The safety man stiffened at the words "office worker." He looked around the tent to the scowling faces. "Look, I hear you," he said. "But the policy is right from the head office, and it's written in stone. I can show you in the policy book." He almost whined. "It's written down."

"Memos are written by men! Men can change."

Acastus appraised the new kid.

"Well," said Pops, "why don't you type up one of your letters and tell them in the head office that these goggles are full of shit and they don't work?" There was a pause in the room as the words sank in.

"Well, I could try," said Acastus. "But I know what they're going to say."

"At least try, Barry," Pops said. "You're our voice to those guys. We depend on you."

"Okay. Okay, I'll try."

The hoarding sighed. Everyone smiled at the safety man. Acastus nodded to the crowd, and then looked at his watch. The plywood tent door slammed.

There was a pause as they waited for someone to pick up the thread of the Toolbox Talk.

"Do you think he'll write that letter?" Dougdoug asked.

"Not a chance."

* * *

"Thanks for backing me up in there." Acastus sneered.

The Safety Nazi had been waiting for Jason. Jason made a show of looking at his watch, mimicking the safety man's escape from the meeting. Then he slowly crossed his arms.

"What are you up to?" Acastus asked.

"What do you mean?"

"When I walked in, you said, 'Don't let Safety catch you.' Don't let Safety catch you doing what?"

Jason glared at the safety man.

"If they're breaking rules, I have to know."

"Do your own bird-dogging," Jason said. He turned, and Acastus grabbed his arm.

"If I catch them, and you know about it...."

Jason shoved his face so close to the Acastus that their hardhats clacked. "You're a little man with a little power. Fuck you."

The crew was at coffee when Jason stormed in. Without waiting for the conversation to subside, he shouted, "The Safety Man heard part of what I said to you about not letting Safety catch you. When I wouldn't give him names, he threatened me. Stash!"

Stash slowly looked up from his coffee.

"Get rid of those glasses. And anybody else. Make sure everything you wear is regulation. Don't give him any excuses."

Jason exited, slamming the door behind him. The room shook with silent vehemence.

"I guess he won't be writing any glasses letter," Pops spoke into the silence.

When the crew climbed into the coker, animated discussions took place on every level of the scaffold. The air in the coker, usually filled with smoke from a dozen cutting torches and welding rods, stayed clean.

"He don't give two shits about safety. He just wants to bully us."

"Those rules don't make any sense. If a safety man has to enforce stupid rules, he should tell his bosses they're stupid rules, and if the bosses won't change the policy, he should quit."

After a time, the men dribbled back to their work stations in groups of two and three. Slowly, level upon level of the scaffolding filled with the comforting smoke and noise of a construction site.

"Watch out!"

"Comin' down! Watch... HEADACHE!"

The universal cry of something falling was taken up. Over and over, HEADACHE! HEADACHE! The sound and feel of something heavy thundering through level after level of crowded scaffolding was warning enough. The thunder rumbled down a hundred-foot drop past layers of scrambling men, men who had nowhere to run.

Bang! Flashing silver in the half-light, the object spun as it hit a hand railing.

Metal hit steel, then ricocheted across the space. Men shrank away, their arms shielding their faces.

A string of safety lights snapped in two. A cascade of sparks, hot glass, and sparkling electrical wires followed.

Men ran. Men shouted. Wide-eyed men hugged steel walls.

Crack! It struck. It was driven almost all the way through the plywood of the bottom scaffold.

The only movement for several seconds was the clatter of dust, falling debris kicked up by the projectile and running boots. Two lightbulbs that were left intact swung in a disjointed circle, making jerky shadows on the walls.

Twenty men looked down from the railings at the silver pipe speared halfway through the plywood scaffolding.

"Everybody okay?" Pops shouted up at the silhouettes.

The shadows shifted as the men solemnly looked at each other. A couple of arms far above waved the all-clear.

"Wow. It missed everybody," Dougdoug said.

Pops shouted up to the heads. "Nobody tells! Nobody tells Safety fuck all!"

And they didn't.

DAY SIX

(Big Mistake)

Gwen Medea got the name Secretary Scary on her first job from a crew of sheet metal workers after she destroyed the career of a tin basher.

For weeks after that job began, Gwen had been the recipient of every form of lurid observation by a fat, forty-year-old chicken farmer turned tinsmith.

Silent and cringing-timid when alone, in front of an audience he became a clown performing for the dubious enjoyment of his fellow construction workers. The man was the embodiment of the reason all oil companies had adopted zero-tolerance harassment policies. Gwen and the other secretaries were amazed that someone had married the idiot and actually had to kiss that mouth.

As the days went by, the comments progressed from off-colour to lurid to downright graphic. Gwen dreaded that it would only be a matter of time before he would no longer be satisfied with simply verbalizing his smut. She had to fight back.

Her chance came on payday. At last coffee, the entire crew crowded around her at the lunch trailer, waiting for their weekly paycheques. With shaking hands, she began handing out the pay envelopes. Gwen had made sure Potty Mouth's envelope was the first one she would hand out. She wanted the entire crew to be close.

As Gwen handed her torturer his paycheque, she held the envelope long enough so that the man looked directly into her eyes.

"Do you know the difference between this cheque and you?"

The sheet metal worker blinked.

"I'd blow this cheque."

Then she released the envelope.

The brutal laughter from the crew made it very plain that they had grown tired of the simpleton's comments, and they immediately sided with the young woman. They were as relieved as she was when she fought back. Like a flock of pigeons, they pecked at him, never letting him forget his humiliation.

Primly, Gwen turned and left the men to their merciless laughter. When she was alone in the washroom, listening to her distant phone ringing, her emotions erupted. She held the scratchy toilet paper to her eyes and mouth as her body shook with fear and release. An hour later, she left the cubicle.

She next gathered the other secretaries and together they went not to the foreman, not to the general foreman, but all the way to the supervisor of the entire jobsite.

The supervisor was not a stupid man. Faced with his entire secretary staff standing over his desk and furiously describing Potty Mouth's antics, and knowing that these ladies could instantly shut his job down if they quit—and make his life miserable in a hundred different ways if they didn't—it was no contest.

Construction workers are usually laid off in groups of threes and fours as a project winds down. But when there's plenty of work left and only one man is let go, there's a reason.

Next morning before work, Potty Mouth's foreman walked up to him and said, "Get your gear, you're done."

They say that his coffee was halfway to his lips and the mug stayed there for half a minute.

Nineteen men sat in that lunch trailer, and not one spoke on his behalf.

Every secretary on that project stamped DO NOT RE-HIRE in

red capital letters on Potty Mouth's personnel file. The same rubber stamp was passed from delicate hand to delicate hand and used over and over to crucify the sheet metal worker. Every one of those files with DO NOT RE-HIRE stamped on the cover went to the Human Resources Departments of all the companies on that Fort McMurray site, and to every union hiring hall. His mouth, and those secretaries, destroyed his career.

Years later, a sheet metal worker told Gwen that Potty Mouth had lost his farm.

"I wonder what happened to his wife?" was Gwen's only comment.

That was then, this is now.

Gwen handed Acastus a coffee. "Two sugars, two creams, right?"

Acastus stared at her. He looked down at the coffee, then back to the secretary standing in front of him.

"I want some information," she said.

Acastus opened his coffee-free hand, waiting.

"You're a knowledgeable guy, Barry. Take me step by step through the harassment policy."

"Why? Were you harassed?"

"I spent my career working with men, and every one of them thought that the answer to all my problems was his penis. I've got good antennae and I have been called a lot of things, and I want to be prepared."

"Why?"

"Because this man meant it."

"What's his name?" Acastus said, setting aside the coffee.

"No."

"Was he one of ours? What'd he say?"

Gwen stared at Acastus stonily. Finally she spoke. "No."

"Well, I can't help you if you're not up-front with me...."

"No."

"Well, I can't help you if—"

"Just answer my question, Barry."

"What's the sexual harassment procedure?"

"Yes, Barry."

"Well, at first, you have to tell the person that his, or her, advances are unacceptable."

"That's not going to happen."

"Why not? You should at least—"

"Barry, no."

"Okay then, you should talk to your supervisor and me as your Human Resources Officer. And we'll either be there when you talk to this person or we'll speak for you."

"That will be four people that know about this."

"Gwen, I'd lose my job if I talk about this."

"It's not your job I'm worried about."

"Gwen, I think your mind is already made up. You're going to let this, this... Hell, I don't even know who we're talking about. But it sounds like you are going to let this asshole slide?"

"This man is dangerous, and I'm not... Barry, if I go ahead and lodge a formal complaint, what happens?"

"We immediately remove this guy from camp. I'm assuming he's in camp?"

Gwen nodded.

"He goes to another camp. Like right now."

"Security? Security will take him?"

"They escort him."

"How many people will know why he's being moved?"

"A couple. The main man will get the memo, and the escorts. Once a written formal complaint is made, an investigation will be started, and a formal report will be written. We'll have to ask a lot of questions. Do you have witnesses?"

Gwen shook her head. "His buddies at the table."

"Well, we'll have to cross that bridge when we come to it. After the report is written, a decision will be made and we'll tell you and this guy what the outcome is."

Gwen stared out the window. "How long?"

"Week, maybe two."

Gwen looked at Acastus. "Thanks for the information, Barry."

Acastus had the distinct impression that he was being dismissed. "We can talk some more... if you like?"

"Thank you, no."

It wasn't his imagination—he was dismissed. Acastus picked up his ever-present clipboard and started to leave. Halfway to the door, he spun around. "I have some pamphlets I'll get for you," he offered.

Gwen's eyes never left the computer screen as she said, "Won't be needing them, Barry. Thank you. Have a good day."

The office was silent but the sound of the door slamming shimmered like the shadow of an echo.

Gwen sat with her hands on her lap, alone in that empty sixty-eight-foot trailer, squinting at the winter sun.

She picked up the phone. "Hi, Angela? Hi, it's Gwen over at Golden and Fliese. Say, I had an incident the other day with one of your guys which kinda left a sour taste." Gwen said his name. "I know he seems like just another loudmouth, but I thought I'd let you know..."

Gwen listened on the line.

"I know he sounds harmless," she continued, "but he's...."

Gwen was silent, listening. The late afternoon sun caught her jaw tightening.

"Angela, do me a favour, please..."

Gwen listened more, her fingers rearranged her mousepad in line with the edge of her desk. There was a brittleness in Gwen's voice when she spoke again. "Angela, Google him. Just Google him. If that doesn't bother you, don't call me back. If you want to do something about what you read there, call me back in ten minutes."

Angela called back in five.

(Email Day Six)

From: Doug
To: Dad

Subject: Fort McMurray

Hi, Dad!

Today two scaffolders got into an argument. One pulled a knife and the other pulled a hammer. They were circling each other when the foremen showed up and stopped the fight. They were separated and then fired.

If you get fired on any one of the refineries in Fort McMurray, your name gets blacklisted with them all. It's something to think about. You could instantly drop from making over a hundred thousand a year to collecting welfare.

Yet some guys still do it.

Lobotomy told me he walked into the local pizza parlour and ordered a meal. They refused to serve him because the last time he was there he caused so much shit they threw him out. Lobotomy said he didn't remember. I believe him.

Pops said, "It's hotter'n two rats screwing in a wool sock."

There was a guy named Scrap Iron who was a bully. He worked on another crew, but every morning when he walked through our trailer he made sure he walked by Stash's table and slammed his fist down on Stash's lunch bag. The guys said that Stash and Scrap Iron had a history.

So this goes on for a couple of days, Scrap Iron walking by Stash, and then slamming his fist into Stash's lunch bag. Finally, after a couple of days of this, Stash disappears for an hour in the refinery's machine shop.

Next morning Scrap Iron walks by Stash and his lunch and like every morning, he slams his palm into the lunch bag, right into six inches of needle-sharp welding rod. It went right through his hand.

Pops' dream is to take a bus trip to Nashville. I asked if once he got there, whether he would get me a date with that singer Sara Evans. He said, "I do believe she is connected, Dougdoug. Besides, I think you are more Minnie Pearl's speed."
I said, "Minnie Pearl? Isn't she dead?"
"That's right."

There was a guy down the hall from me in the construction camp who was found hoarding 276 of those breakfast-sized boxes of Raisin Bran. So they fired him for theft. The comment was, "I dunno, he seemed like a pretty regular guy."

Pops describes the way The Safety Nazi walks is like an arrogant penguin. He doesn't know the borderline between safety and coercion, or as the guys say, he's "co-worsting" us. He seems to forget that almost all these tradesmen have transferable skills and if he fires them from this project, they'll get on the cell phone and be heading for another job by the time they leave McMurray.

Like Pops said, "You get the union you deserve." Sometimes after the weekly Safety Meeting, and being on that half hour receiving end of one of his "talks," I think the Safety Nazi would really like to be issued a German shepherd.

Dougdoug

DAY SEVEN

(Morning Warning)

The lunch trailer went still as the foreman and The Safety Nazi entered. Jason looked down at his clipboard.

He nodded to a skinny welder. "Lobotomy, I suppose you missed yesterday because you were sick."

"Naw," Lobotomy shot back, "yesterday I was drunk. Today I'm sick." He laughed loudly as he looked around the room, gauging the joke's effect.

"You and the Steward, after the meeting, in my office."

The laughter died.

When Lobotomy was an apprentice, he began signing his name Lobo, which means wolf. He quickly regretted it because immediately everyone else started adding "tomy" to anything he signed.

Lobotomy was a skinny young man of many enthusiasms and body tics and few talents. If Lobotomy were a car, he would be in a lifelong demolition derby, careening from one near-wreck to another.

Lobotomy has been a timid punk rocker, a tattoo and body piercing addict, a skinhead terrified of violence. Recently Lobotomy reincarnated himself one last time. People said that Lobotomy was in the process of slow-motion suicide. When Lobotomy heard those words, in the drug-crusted recesses of his brain an alarm bell finally clapped. It dawned on him that as cruel as the comment was, it was the truth. That's when Lobotomy found Jesus.

But when Lobotomy found Jesus, he didn't actually pray. He felt that if he told enough people he'd found the Lord, well, that was good enough.

"Okay, guys," Jason said, "this morning's Toolbox Talk is about ladders. Ladders are to be positioned so that there are at least two rungs above the level where it's leaning. Ladders are for people only. Do not carry equipment up a ladder. Put the equipment in a canvas bag and pull it up by rope."

The Safety Nazi pointed to Lobotomy, who was waving his hand excitedly. Jason nodded in his direction.

"Do you know why it's unlucky to walk under ladders?" Lobotomy asked.

"Yes," Jason said, "I think I do, but I'm sure you're itching to tell us."

"Well, that's the Holy Trinity. Each point is the Holy Trinity."

The room went silent. Lobotomy smiled angelically. After the extended silence stretched too long, Lobotomy continued.

"Each point the ladder makes is The Father, The Son, and The Holy Ghost. You can't break the trinity."

The room remained silent.

"After the meeting," Jason said, "you and the steward, my office."

"No, really," Lobotomy blurted.

"Yes, really. My office, after the meeting."

Stash looked at the Safety Nazi and said in a theatrical tone, "Hey, Barry, is it true you fired the maintenance foreman's daughter?"

Acastus turned purple. "She wasn't the maintenance foreman's daughter, and she walked through a red ribbon. Let's move on."

Stash was referring to a beautiful young university student who had been hired to work the summer with the crew cleaning the lunchrooms. At the end of the orientation, she met her new foreman. She signed the papers, and she walked out the lunchroom door following three other new workers. All four new-hires walked under a red ribbon. A red ribbon on a construction site is the ultimate "no go" area. There is imminent danger within any area designated by a red ribbon. You do not take a shortcut over or under red ribbons; you go around.

The Safety Nazi fired all four of them on the spot.

The young beauty's sobs were stifled as she left the office door. The Safety Nazi's eyes burned bright. Pops, who, as union steward, had to witness this casual cruelty, turned to the smirking Safety Nazi and said, "You know who her father is?"

"I don't give a shit," Acastus replied, but his eye was twitching slightly.

Within an hour of the young beauty's firing, the Safety Nazi's truck was taken into the refinery's maintenance shop for repairs, and no temporary vehicle was issued. He immediately had to either walk or borrow someone else's truck.

Acastus went back to his office, where he was met by a red tape of his own across his doorway bearing a tag that stated the room was being painted. He wasn't issued a new office. He now had to share a desk with a secretary. Then his computer suddenly needed debugging. He now had to write his never-ending reports on the secretary's computer. The secretary wasn't happy.

The crew would put money on the fact that the Safety Nazi gives a shit now.

"Let's get back to ladders," Jason said, smiling at Acastus' discomfort.

So the Toolbox Talk went on. The safety bulletin was finally read, the day's work allocated, "hot work" permits issued for anyone welding or cutting with torches. Lobotomy was in Jason's office with Pops, the job steward, waiting to be given a verbal warning about missing time and not phoning in.

The day had started.

"Is that true about the Trinity?" Pops asked Jason when he arrived.

"Hell," Jason said over his shoulder as he entered his office "I dunno, but I'll bet we're going to find out."

DAY EIGHT

(Party)

The boilermaker crew was staying on the second floor of a local hotel. Much to the construction workers' delight, a troupe of traveling strippers moved in directly across the hall.

Every evening after dropping their laundry on the tiny white round stage downstairs, the pole dancers would come back to their hotel room and sit in their silk housecoats for hours watching soaps on TV, doing crossword puzzles, or talking on the telephone. Who knew that being the sexual fantasy of a couple of dozen semi-housebroken orangutans could be so boring?

The boilermakers, however, were in construction-worker heaven. In Fort McMurray, there are thousand-man construction camps with fewer than fifty women in them. Any woman is a rarity, and a woman for whom clothing is non-compulsory is a dream.

The four welders left their door open across the hall from the working girls, and whenever the ladies' door would open, immediately four heads would peek out of the boilermaker's door to deliver a chorus of:

"Morning."

"Morning."

"Morning."

"Wanna get naked?"

The ladies would always giggle and wave and it was an uncomplicated encounter—in one form or another, everyone in that hallway was busy hurrying off to work.

This went on for several days until there was a change in work rotation for the welders. For the next two weeks, they had their evenings off. It was time to make their move.

First things first: the men topped up their hoard of fresh liquor. They even went so far as to buy white wine, because isn't that what women drink? They stacked cases of beer in clear sight of the door as a lure, liquid breadcrumbs for the bunnies across the hall.

"What are you doing?"

"Throwing out these cups."

"Why?"

"Women get really upset when they see cups half-filled with tobacco juice lying all over the place."

"I guess."

They all washed, sort of. They changed into their cleanest blue jeans, checked their breath, and then, with an anxious last look around and a final kick to the underwear under the sofa, the boilermakers went a-courting.

The welders screwed their courage to the sticking place, marched across the hall, and pounded on the strippers' door. When a vision of loveliness dressed only in a housecoat opened the door, a bottle of ice-cold Pinot Grigio was thrust towards her cleavage.

"You guys wanna have a drink?"

"With us?" another welder quickly added.

They might as well have added, "Oh gawd, please, please, please."

The vision motioned them to stop. The door closed, leaving the men to fidget in the hallway. After twenty minutes of giggling, thumping, and muffled conversation, the door reopened. From the room emerged four of the loveliest little girls the men had ever seen. They were dressed, scrubbed, their hair pulled up into ponytails, their teeth white and regular. They looked like a team of cheerleaders.

The men were just happy they had a pulse.

The men sat, their backs stiff, aware of the warmth of the women beside them. They had all seen the ladies buck naked but this was different: the girls were clothed, they smiled, and talked. One of them had even had dimples. It came as a shock that these young women sitting beside them were people—people with thoughts, opinions, and actual feelings.

The men felt a little crestfallen and more than a little ashamed. In unison, the men raised their beers to their lips and took a long, desperate drink. The men had half-expected an instant Roman orgy, but what they were getting was a meet-the-teacher interview.

The littlest lady with the friendly dimples watched the welders take a long pull on their drinks. Her eyes shot a quick glance at the other women and then stopped at two fishing poles in the corner of the hotel room. She asked, "Who's the fisherman?"

He was taller than the others, with jet black hair and what looked like a four-day beard growth. He could have been Spanish, Italian, or Mexican; he was Métis.

"I am."

"What's that like?"

He shrugged. "It's... fishing."

"Well, how do you do it?"

"You've never fished?"

The women looked at each other, and back to the fisherman. "No."

The fisherman sat on the edge of the hotel's sofa, twisting his beer in his hands. "Well, first thing you gotta do is make sure there's fish where you set up. Ask around, find out what type of fish there is. Ah, in your spot."

"How do you do that?"

"Ask at the place they sell fishing gear. If they don't know, nobody does. Once you find out what fish you are looking for get the proper..."

"I thought we were going to get naked," blurted out the smallest of the welders.

"...bait."

Dimples looked directly at the small man. "It's Sunday." She glared at him. "C'mon, ladies, time's up." She stood.

"No, no, no!" said the big welder. "It's okay! Stay. Please." He looked at the smallest man, who had said the unthinkable. "If these ladies leave, so will you." The other welders looked at the man. "And they'll be allowed to come back."

The small man took a quick pull at the beer, glaring at the rug.

The women settled primly back in their seats, their eyes flickering at the door once or twice.

The fisherman got up and opened the door to the hallway. The ladies settled back into their chairs. Dimples smiled at the fisherman. "So you picked your spot, what then?"

The fisherman looked at her. "Oh, oh. Aw, well... then you look for bait. And you make sure the hook is small enough so the fish can swallow it."

The fisherman talked about setting the hook, making the proper knot, getting the right bait, and on and on. The room fell silent as the women and the welders listened to a man who truly loved his hobby. Late into the night, the man's voice resonated, interspersed with the softer voices of women asking questions. Fishing rods were handed around, lures were discussed, and the wicker creel for storing fish was tried on.

North of Fort McMurray, past the massive Syncrude and Suncor refineries, there's a construction camp just to the west of Highway 63 called Barge Landing. If you drive past that thousand-man camp and continue down the hill, you'll find a huge dock where barges land and off-load equipment from the Athabasca River.

When dawn came to Barge Landing, it found a tall Métis fisherman with four giggling ladies standing around him trying out their newly learned casting technique. Off to the side were three welders making love to their last beers and staring into the bonfire.

The next day, every man in the crew went out and bought fishing rods.

DAY NINE

(Oxygen Content)

CRACKLE.
  "Jay, are you there?"
CRACKLE.
  "Come in, Jay."
CRACKLE.
  "What's up?"
CRACKLE.
  "Jay, the Hole-Watch won't let us enter the coker because she says that her oxygen reading is off the charts."
CRACKLE.
  "Well it should be reading between nineteen-and-a-half and twenty-three percent. What's the reading that she is getting?"
CRACKLE.
  "The Hole-Watch says it's eighty-point-two."
CRACKLE.
CRACKLE.
  "Turn the monitor right-side up. What's the oxygen reading now?"
CRACKLE.
CRACKLE.
  "What's it read now?"
  "Twenty-point-eight percent oxygen."

CRACKLE.
"I think you can go into the coker now."

(*Emeralds and Snakes*)

"Ouch! Ouchouchouch!"

"What happened?" the old man asked.

"Banged my finger," I said, examining my ungloved finger.

Pops joined in the scrutiny. "Hmm, well, could be worse," he said. "Could be my finger. So let's take a break."

I sat back, shaking my head. "I can keep up, if that's what you're thinking," I said, every muscle, bone, and joint in my sixty-four-year-old body aching. "I just want this smoke to clear."

Pops smiled.

The string of safety lights jerked the shadows inside the coal-dark stomach of the coker. No colours escaped the all-pervasive black. I took off my glove to examine my finger. The age lines on the back of my hand were black and deep in the brittle light. Our eyes were pits, all expression hidden in the shadows.

The work had been going well. We made a good pair. Pops and I had about a gazillion years' experience between us. Pops had said several times that we should be finished our job well before the end of the shift and we could afford to steal a moment.

"How long you been a boilermaker, Pops?"

Pops looked away and was silent. I shifted my position. He spoke as he looked towards the wall, reliving long-forgotten pain.

"Too long."

"No, really."

"I had my twenty-second birthday working night shift, welding in a bloody gravel pit. All night long, every time I shut off that portable welder to refuel or take a break, I could hear music coming from a party at a golf club across the highway. You could just hear the sound, kinda shimmering, coming and fading away at the same time. Once,

I heard a woman's voice laughing. I stood in that fucking gravel pit, listening."

Pops replaced his glove, worked his fingers once or twice, and continued.

"I imagined her. To me, she had blonde hair, wide-set eyes, smiling face, low-cut red dress, no bra, and nipples you could hang your coat on."

"You should have gone and joined them."

"Yeah, I could just see it. Me and my welding leathers and helmet, walking into some rich man's party."

"That'd work."

"I never felt so low. Here I was pouring welding rod, swatting mosquitoes, listening to rich folks having a party. Happy bloody birthday. Well, anyway, that job fixing equipment in a gravel pit got me a job in Brazil."

"Brazil?"

"I was working for that same construction company when they got a job punching a road through the jungle. I had just gotten married, and I figured I'd make a whole lot of money fast. I wanted to be one of those laughing voices at that golf club."

"What, this ain't good enough for you?" I smiled, indicating the black cavern surrounding us.

"Being this dung beetle? No. My wife didn't want me to go. She cried, but I wanted to make that money. I wanted to go... make a fortune... so bad. I'd do anything to get out of that gravel pit. So I ended up driving a Cat through the jungle."

"Gravel pit to jungle," I said, my voice flat.

"It was a D9 Cat, the biggest... well, at the time, the biggest Caterpillar Tractor they made. I started off trying to knock down the trees like we do here in Canada: raise the blade and beat hell out of the tree. But I learned real fast that if you did that, snakes would fall out of the trees and onto you. We had huge umbrellas attached to the Cats for shade. When a snake fell and hit the umbrella, you'd hear this thump and see a shadow squiggling down off the umbrella.

Then the snake would drop onto the hood, the tracks, or the back of your seat. I spent months driving tractor while looking around my feet and ass for snakes."

"Poisonous?"

"Real poisonous. They told us, 'Red and black, you're all right Jack. Red and yellow will kill a fellow.'"

"Anybody got bit?"

"I had one try to bite me. I'm working away, minding my own business and I felt something bumping against my pantleg. I look down, and there's this snake trying to bite me. But his mouth was too small. I guess if he could have got ahold of the skin between my thumb and finger, he would have killed me, but he was too small. You could tell he was mad. Just what I wanted: a pissed-off poisonous snake on my leg. So the new camp they dug out of the jungle is infested with snakes: under your bed, in the shower stalls, everywhere. The camp manager goes out and hires local kids to kill the snakes."

"What did he pay them?"

"Five cents each, and that was the problem. These kids didn't see five cents a year. Some of those kids were bringing in twenty, thirty snakes at a shot, then going home and getting robbed by the men in their village."

"Wonderful thing, civilization."

"So pretty soon, there's no more snakes, none. The camp manager thanks the kids, and then tells them to fuck off."

I leaned closer to hear the story better.

"A couple of days after the kids were fired, the camp is infested with more snakes than before. It was so bad, we had to sleep in the trucks. Shit, there were snakes everywhere. It turns out, the kids the guy fired went back out into the jungle, caught as many snakes as they could carry, and threw them all over the fence into the compound."

The only white in the coker was Pops' smile.

"So the camp manager decides, screw the kids, he's going to get something that'll kill snakes and not turn on him. So he gets in a planeload of pigs. Pigs eat snakes. Pretty soon we got this herd of pigs wandering around, killing all the snakes in the compound. In a couple of hours, all the snakes are gone."

"That's it?"

"Oh no, it gets better! To a native, a pig is riches. Whole villages can only afford one pig and our small crew had a dozen. Within a day, we got all the natives from miles around standing outside the fence, staring at those pigs. They just stood there, nothing moving but their eyes, staring at all those pigs."

I shifted, trying to stay warm.

"First night the pigs came was real quiet. Next night, in the middle of the night, all you heard was 'creekcreekcreek' as one pig after another got stolen. And some of those pigs were big. It got so that we only had one pig left, and he had to be kept locked up in a pen. The camp manager had to go out and hire a local to guard that pig."

"I'll bet he loved doing that."

"Oh, he was pissed."

"Finally," said Pops, "the camp manager had to rehire those kids to catch the snakes, but this time, he only paid them a salary. So we got to know the local villagers through those kids. One day, one of the kids brings in an emerald. Every white guy in the place drops what he's working on and rushes over. First the kids, and then the entire village started bringing in more and more emeralds. Then the local mining company comes over to the boss and tells him to get us to stop. We were driving up the prices."

"Did you get lots of emeralds?"

"My pockets were full. So where was I? Oh, these mining guys, they were hard cases, one up from banditos. They told us not to buy from the villagers, and they told the villagers not to sell to us or their wives and children would disappear."

I stared at the older man.

"Just before the job was over," he continued, "we rented a boat, one of those thirty-foot canoes with an outboard. And we go about forty miles upriver, right into the place the emeralds were dug."

"I thought the villagers did it."

"We thought so, too. But turns out, the villagers were the first in a long line of middlemen. These guys we visited upstream were the guys that actually dug them out. We were buying emeralds for fifty, sixty bucks each."

I let out a low whistle.

"The day after we got back to camp, we shipped out. We heard the mining company guys were real pissed we had trespassed all over their claim."

"Makes you wonder what they did to the guys that sold you the emeralds."

Pops shrugged, thought a moment, then said, "Well, they were employees…. Anyway, we get into Rio and we're going to fly up to Canada the next morning. Everybody takes their emeralds and jams them into toothpaste tubes and hollowed-out soap bars. They even took condoms, filled them with emeralds, and swallowed them. I get back to Winnipeg, and my apartment was empty. All there was, was an empty wine bottle. My wife had taken off with some guy. She left me a note, though. It was from some lawyer. She charged me with desertion."

"Ouch."

"I'm not much of a letter-writer," said Pops.

"What do you mean?"

"I didn't write."

"How long were you gone?"

"Nine months."

"Nine months and you didn't write one letter? Did you phone her?"

"We were in the jungle."

"At least you had the money from the emeralds."

"After the lawyer's bill, it was pretty much a saw-off. I got to keep a couple of stones but everything else went to her and the two lawyers."

I lowered my head but my eyes studied the old man.

"You know," he said, "about a year after I got back, there was a party out at that same golf club. Most of the club was off-limits, but I walked in that hall, all dressed up and waiting. I thought that club, well, you know, owed me."

Pops was quiet for a while.

"Pretty cheesy, actually. A lot of fake wood and attitude."

DAY TEN

(*Lobotomy's Phone In*)

"Golden and Fliese, Gwen Medea speaking."
"Hi, Gwen."
"Hi, Lobotomy."
"I'm not coming in today."
"I'm shocked."
"No, really, the Doc says that I have a touch of narcolepsy and I have to stay home and get some sleep."
"Really."
"Yeah, really. I'll be in tomorrow."
"I'll bate my breath. Get that doctor of yours to sign a note because you'll need it to get back on site."
"Okay, Maybe I'll be in later today."
"Miraculous recovery there, Lobotomy."
"I'm a quick recoverer."
"Yep."

(*That Loud Man*)

"I had been working on radar sites in the Canadian Arctic for four-month stretches," I said. "Just flew from site to site to site, finishing one job and then packing up and off to another one. Never saw anybody, just worked. I knew after a hundred days, I sort of went squirrelly."

I looked across the table at the silent man. The late-afternoon sun came sideways through the empty dining hall at Borealis camp. The sun cut crags in his neck, and his forehead looked like a pack of wieners. The man remained silent as I continued.

"Bushed is like walking through fog. When I finally got back home, it was a shock. All that noise, jerky people, flashing lights, and traffic. Cars coming so close you could touch them. And fast? Man, they were fast. Scared the hell out of me.

"On the way home from the airport, we stopped to do some shopping. The mall was torture. My wife would meet me at the airport and say something innocent, like 'How was your flight?' but she sounded like she was screaming 'I hate you!' and I winced."

The man across from me nodded, and took a sip of his coffee. His eyes stayed on me.

"When my kids were small, I was gone. I had to pay off the mortgage, keep the wife and kids in Big Macs. My youngest daughter once told me that she doesn't remember me until she went to school. You know, that's the first five years of her life, gone. I missed them all.

"One year, I remember being away for eight months.

"That's the year the duffel bag never left the foot of the bed. Any conversations I had with my wife boiled down to grunts and money. I'd come home, we'd have a honeymoon for three days, then one day we'd wake up and we were strangers.

"One night, I just got home for Christmas. It was so cold that the snow crunched like biting into a fresh apple. But that Christmas turkey smelled so rich and it was so warm. We were a family, you know? A family."

There was a clatter and a murmur from the camp kitchen behind us. Both of us looked up as a skinny kitchen staff member in starched whites walked by carrying a mop. After a while, I returned to my story.

"My wife was facing me, and the kids were sitting in their booster chairs on either side. My wife had made herself and the kids up, because Daddy was home. My kids and house were always spotless. I'll give her that—my ex was a good housekeeper. Their hair was all up in spiky little ponytails. Faces were shiny and scrubbed. My daughters all wore matching white sweaters and little kilts, red and black kilts with white stockings and black shiny shoes. Daddy was home.

"I'm babbling away to the wife. Happier than shit to be home, and a little voice came from one of my kids.

"'When is that loud man going home?'

"'When is that loud man going home.' I don't think I've ever been hurt like that by anybody. I was just numb. My wife gave me shit about not making a fuss about the way the kids were dressed. But I just sat there playing with the food.

"'When is that loud man going home?'

"I tried to stick around a little longer, but you know, really, they didn't need me. Oh, they needed the money. But they had their own lives. They grew up fast.

"Everybody got used to me being away, and you know what? So did I.

"When I did take a job in the city, my kids treated me like an uncle. Still do.

"The more friends you have, the happier you are. Us boomers? We don't have friends. We slop back and forth across Canada like water in a pan, never talking to anybody. Our long-term relationships are a weekend.

"That year up in the Arctic, I made over a hundred grand. That's good money even now. The money's gone, and my kids? They have their own families.

"Their husbands don't travel."

The old construction worker and I mumbled a couple of times but the fire in the conversation had flickered out. Awkwardly he rose. I raised my coffee cup in mute salute. He glanced at me, his eyes lingering an extra second or two. He nodded, then gathered his tray of dishes and walked away down the long row of dining hall tables.

    I sat alone in that six-hundred-man dining hall, listening to the kitchen staff rattling in the distance. I held my coffee cup and stared over it south towards the highway. The late-winter afternoon sun made my eyes water.

DAY ELEVEN

(C6)

"Rick, I got a job for you. You and Teddy are welding the trays in C6."

"What level?"

"All the way."

"All the way?"

"Yeah, and don't you be like that pipefitter that got halfway up yesterday then told everyone he had to take a piss. They're still looking for him."

"All the way?"

"Look on the bright side: nobody will bother you."

Refinery towers are essentially pressure cookers that use heat and gravity to separate crude oil into various thicknesses. First, the crude is steam-heated. Then the fumes generated by the boiling oil separate as it wafts upwards. The lighter the fumes, the tinier the molecules, and the higher they float. The heavier the droplets of crude, the lower in the tower they will re-form, until the top of the tower collects light naphtha and the bottom of the tower drains off something akin to black toffee.

The inside of a tower looks like an acoustic testing chamber, except an acoustic test chamber has foam rubber sticking into the compartment. Jutting into a refinery tower, on the other hand, are razor-sharp steel shards. Workers crawling around inside a tower wear knee pads, elbow pads, hardhats, and coveralls that are sure to be ripped. After a couple of days of creeping through a tower, the worker is guaranteed to end up cut and bruised, his clothes in shreds. The interior of a refinery tower has all the comforts of an inside-out porcupine.

I could hear Teddy puffing below as I climbed. I stared at the next yellow rung, the next placement of my gloved hand. I didn't look down, up, left, right, or anywhere. I looked at the next rung, only the next rung.

My hands squeezed the yellow-painted rung as I felt the tower sway in the wind. I didn't think that was possible. Teddy chuffed and spat below. The next level was ten feet above my hands. I glanced down, and promised myself not to do that again. Teddy, with his battered hardhat and his massive arms, was below me, the crane the size of a tiny, perfect children's toy far below him. At this distance, the crane's massive diesel motor was a hum, almost a moan. Teddy's breathing was laboured. Teddy looked like a thumb, a thumb with a hardhat. He wasn't my height, but he was all of my weight and half again. His beautiful, ice-blue Slavic eyes crinkled and cried in the McMurray cold. With his red face, he looked like a quiet and sad Santa Claus.

The closest anyone could get to pronouncing Teddy's full name was something close to "Teddy Half-a-can-a-gas." Everybody just called him Teddy, Teddy Consonants, or Teddy Alphabets.

Teddy had been a twenty-year-old welder in the Gdansk shipyards when the German army attacked Poland at the start of World War II. When the invasion started, Teddy had just enough time to drop his tools, change his clothes, join the Polish army, and get captured by the Germans. "Pretty quick" was his only comment.

After being captured, Teddy spent years twiddling his thumbs in a German POW camp. The camp was eventually captured by the advancing Russian army. The Russians greeted the Poles by putting them in an internment camp where Teddy and several thousand POWs sat for another couple of years.

Once the war was over, the Russians opened the doors of the camp and thousands of men, including Teddy, were let go. Teddy walked across Poland, Germany, France, and into Belgium, where he registered with the Allies as a refugee. The Allies gave Teddy clothes and food and put him in an Displaced Person Camp, where he sat for another year.

By the time he arrived in Canada, Teddy was pushing thirty. On a landing halfway up, I called a time-out.

Teddy smiled. "High, eh?"

"High," I agreed. Pointing to the steam clouds on the horizon, I said, "McMurray's over there, down in the valley. Suncor's over there. And that cloud, back over there, should be CNRL."

"I was at Firebag last year," Teddy said. "It's over the horizon, almost a hundred klicks. I was in a refinery once in California."

I looked at Teddy. I was about to ask him how he got to California when he continued.

"It's got a twenty-five-storey tower there. Looks like a bullet. When we climbed up it, for the first ten stories we took a man-lift—you know, a vertical conveyor. No cage, no protection, no nothing. We stood on those foot angles, hung onto the hand-grab, and watched the refinery get smaller. If you let go, you died. When I stepped off the man-lift on the tenth floor, there's an outline of a body painted on the cement, and a name and a date stenciled beside it. Makes you look up. Where that guy fell from was where I was going.

"I climbed the rest of that hundred and fifty feet by ladder. And do you know what I saw when I got to the top?"

"What?" I said.

"You'll never guess. Not in a hundred years."

"Okay I'll bite. What?"

"A bicycle."

I stared at Teddy.

"Somebody stole his foreman's bicycle and chained it to the top railing of the highest tower there."

"How'd they get it up there?" I asked.

"Dunno. When I asked, all they did was smile."

"Putting a bicycle on your back and climbing the ladder would be awkward."

"Pretty dangerous too," Teddy said. "What if it got caught in the ladder's cage?"

"A crane would never reach two hundred and fifty feet, and besides, it would attract too much attention."

Teddy nodded in silent agreement.

Teddy and I thought for a while, our minds far south to another, far warmer refinery.

"Rope," Teddy said. "Musta been a rope. One guy climbing up, lowering the rope to another guy, pulling the bike up to the next landing, leapfrogging all the way."

"Would have taken a couple of men hours." I smiled.

"In the dark," Teddy said.

"Yeah, night shift." I smiled.

"Why would someone pull a bicycle to the top of an over-two-hundred-foot tower?" I mused. "The foreman must have been a real prick."

Teddy smiled. "Naw. They did it... they did it because they could." We resumed climbing. The swaying increased.

Climbing the ladder of a tower, you find out real quick if you have a fear of heights. The entire climb I never spat, because I didn't have any spit.

I stared at the chipped and worn yellow paint of the next rung of the ladder. What if I slip? What if I have a heart attack up here? What if Teddy does? Can I go on? What if I slip and fall into the cage with its steel bars like dull knives? What if? What if?

After a half hour of climbing hand over hand, I lay on the

topmost steel grating. Nothing was above Teddy and me. We had made the climb, we were there, twenty-five stories. Off in the distance, the tops of other towers from other refineries dotted the landscape. Far to the south, you could see the valley where Fort McMurray is. We had done it.

Teddy gave me a puzzled look. He looked around the steel grating, and into the man-way opening where we were to work. He asked,

"Did you bring the grinder?"

(Email Day Eleven)

From: Doug
To: Dad
Subject: Ft. McMurray

Hi, Dad!

Three of us guys were working inside the coker and while we were waiting for a "lift" from the crane, we started talking hockey. One of the guys mentioned that the local team was playing the Okotoks Oilers.
The third guy, appropriately nicknamed Lobotomy, says,
"The Okotoks Oilers, where are they from?"

The job's progressing. For the first few days, all we did was load in a lot of equipment into and around the coker. Mostly, us apprentices and labourers just set up the mechanics and welders so they can do their jobs.

The guys tell me that the sand in Fort McMurray has a hardness factor of seven, ten being diamonds. So when there's anything that moves, shakes, flows, or tumbles the sand, it wears away real fast.

The water and sand mix is so abrasive, all those shiny new-looking vessels that you see in the pictures? They're wearing away from the inside out. And I mean fast.

So during a shutdown, a lot of the welders will stand on scaffolding and just weld on the inside of the coker. Pad welding, they call it. For hour after hour, all they do is weld, huge shiny squares of weld. Just building up what the sand has worn away. It sounds boring, and it is, but if ever they miss a spot and the steel gets worn so much that there's a hole blown in the side of the vessel, you want to be someplace where you can say, "What was that?"

Pops told me this:

"We were working at the smelter in Ontario, on a smokestack. So one day we're going up the man-lift inside the stack, and about halfway up there's a power failure; everything shuts down. So we're sittin' there, about ten or fifteen stories up, inside this stack, for about two and a half hours. Its blackern' coal.

"Finally Carnage says, 'Enough of this shit. I'm sliding down.' So he grabs the cable and goes over the side, and starts to slide down the hundred feet or so to the bottom. About halfway down, he hits a patch of grease on the cable, and starts to free-fall.

"When he finally got down, his gloves were all ripped, his boots were worn through, his pants were all torn. Carnage walked away from that cable a changed man."

Two-Tall phoned in to Secretary Scary and said he couldn't come in. Something about a preliminary hearing.

Mongo and Shaky decided that they didn't want to live in camp, so they rented a room at a hotel in McMurray. Mongo was working days and Shaky was working nights, so though they shared the room, they never saw each other.

One morning when Mongo went off to work, he left Shaky a present.

A whole package of firecrackers under the toaster with the wick inside the toaster, and wrapped around an element. Shaky came home for breakfast, put the bread in the toaster, and he settled back to enjoy his morning coffee... but not for long.

Mongo said, "I only took welfare twice in my whole life! Once for fourteen years and the other time for twelve years."

Pops should be retired. Four marriages and a couple of liaisons means he'll never get to sit on that front porch and watch the grandchildren play. The way I figure, if he did retire Pops would only get a quarter of his pension. The rest would go to his exes.

There's a lot of these old guys like Pops in Fort McMurray, Dad. You see them everywhere, old men, still trying to do the work of thirty- and forty-year-olds. They are way past retirement age but they're everywhere. I can only imagine that they have either screwed up their personal lives, financial lives, or they are just trying to work one more shutdown, one more year, or get just one more payday under their belt before they pull the pin. The other day, somebody said a crew of ironworkers averaged fifty-seven years old. Some of the construction camps here in Fort McMurray are, by and large, populated by old men.

I'll email again in a couple more days.

Doug.

DAY TWELVE

(Dinosaur Farts)

Dougdoug stood inside the coker, lumps of black coke oozing into his already oil-soaked gloves. "You know, Rick, I've been thinking."

"Always dangerous. What about?"

"Oil."

"Oil? Of course! We're surrounded by it. If your fingers weren't so dirty, you'd be picking your nose with it. Why wouldn't you?"

"You making fun of me?"

"What ever gave you that idea?"

"Makes you wonder. Where did this oil come from? What was oil before it died?"

"Funny." I tapped my hardhat in an exaggerated way, as if I had just discovered a great truth. "That's exactly what I was thinking."

"Screw you."

I smiled. "Pond scum."

"Then why doesn't oil smell like pond scum?"

"Maybe trees in a forest. Shit, I dunno! Why don't you ask one of the engineers?"

"Then why doesn't oil smell like a forest?"

"Maybe that's what the forests smelled like back in the Jurassic."

"Naw, Rick. I've smelled forests before. This smells like... oil."

"No shit!" I laughed. "Slow down thar, Wild Thang."

"Well, you know what I mean."

"Well, if I squeezed the shit out of you for a couple of million years, I'll bet you wouldn't smell the same."

"Maybe that's why."

"Huh?"

"Well, all the trees that this comes from are crushed together, so that's why the smell is stronger."

"I had a girlfriend that smelled," I muttered.

"If we diluted the oil... I wonder if that's what the forest would smell like? I mean, in the Cretaceous?"

"Everything spices," I said. "Jeez. Everything."

"Does oil from Texas smell different from the oil from Saudi Arabia?" Dougdoug asked.

"Whole house would stink."

"Gas smells kinda neat. In small doses, I mean."

"And fart! Holy shit, I'd have to staple down the sheets."

"The trees were different millions of years ago. Maybe that's why it smells the way it does," Dougdoug said pensively.

"What? I thought we were talking about my ex-girlfriend."

"Oil."

"Jeez, kid, stay with the topic!"

"Everything was different back with the dinosaurs. The percentage of oxygen in the air, the plants, maybe that's why oil smells the way it does."

"It probably smelled the same as now," I said.

"I don't think so. I read a story about Pearl S. Buck when she was a missionary in China. She said that the worst part after years in China and coming back to America was getting used to the smell of her family."

"So?"

"She said all that butter and meat made us North Americans stink. It's what you eat that determines how you smell."

I smelled my shirt. Dougie smelt his armpit.

"Oil," we said in unison.

Turning to the young man I said, "Dinosaurs probably smelled like alligators, or snakes."

"If you think about it," Dougdoug said, "when you run your truck, the exhaust you smell is probably the smell of ancient pond scum burning."

"Or dinosaur farts."

"I don't think so, Rance."

"Dinosaurs farted! They find petrified dinosaur shit all over the place. If cows fart, so did dinosaurs."

"There weren't that many of them."

"Bullshit! There were millions of them. You know, for a smart kid, you don't watch the Discovery Channel much."

"Not as many as there were trees in the forests."

"Dinosaurs farted!" I actually started to believe myself.

"How do you know?"

"Can't a guy just know?"

Dougdoug and I were quiet for a moment, reloading.

"Okay," Dougdoug said, "maybe dinosaurs farted."

"See! I knew you were a smart kid."

"But you know what? I think the dinosaurs are killing us."

"Dougie, what I just said about you being smart? I take that back."

"No, really! The carbon dioxide in the atmosphere is caused by burning fossil fuels, so actually it's the dinosaurs that are killing humans."

I looked at Dougdoug.

Dougie looked back at me.

"Why couldn't I get an apprentice that only had Grade Nine? I'm sure they've got them."

Dougdoug ignored me and continued. "Another thing. When I said that I wondered what oil was before it died, I guess 'died' was too strong a word."

"How so?"

"'Changed form' is a better way to say it. Oil changed from a solid vegetable to a liquid, and now we're changing the liquid into a gas. A gas that's ruining the atmosphere, causing global warming, and will eventually poison the air."

"Like dinosaur farts?"

DAY THIRTEEN

(Redoing the Job Hazard Card)

"Okay, you three. Redo this."

"What, you don't like our Pre-Job Work Card?"

"Redo it."

"Aw, Jay..."

"Redo it."

"You don't like the Hazards we put down?"

"Even if you are a Newfie, working with 'Off-Islanders' does not constitute a work risk."

"Depends."

"And use your own names."

"Hm."

"So who's Separation?"

"Too Tall, he's from Quebec."

"That makes Dougdoug... let me guess, Alienation?"

"Yep, Red Deer."

"And who's... the Middle East?"

"Rick. He's from Winnipeg."

DAY FOURTEEN

(Frozen Hams)

"You tell the guys, Pops, no more walking out early. At 5 PM we start to clean up. Return all tools to the tool crib. At 5:15, we leave for the bus. The bus picks us up, and we're driven to the camp. You're the shop steward, you tell the men that they're getting paid until 5:30. No more standing around the bus stop at five like Stash and Too Tall did yesterday."

"Okay then, how come the non-union assholes always get the early bus?"

"They have a different schedule, and why do you care?"

"'Cause they all laugh at us. We play by the rules, and they laugh at us when we do."

"Just don't look at them. When their bus drives by, turn your back to them. Jeez, Pops, tell the guys to grow up!" The foreman's boots clattered on the iron grating.

This fight wasn't finished, Pops grumbled. Over the years he had acquired a good understanding about how groups react to situations. So, Pops thought, you just have to create a situation.

Fort McMurray moves on buses. Twenty-four hours a day, seven days a week, buses pick up and drop off workers either at work or back home at the end of the shift. Thirty-bus convoys are not uncommon and one-way bus journeys can run as far as a hundred kilometres.

Highway 63 can have twenty-kilometre traffic jams. Buses play a vital part in keeping McMurray running efficiently. The smooth operation of the buses dovetails into the smooth operation of the refineries. If the bus arrives minutes late or minutes early, it is noticed.

Like ants on an oak tree, the buses climb from the city of Fort McMurray in the grass thirty kilometres all the way up to Suncor, then Syncrude, sitting like silver gatekeepers on either side of the highway. Suncor can't be seen from the road, but the bus traffic looks right down on Syncrude as they pass.

After Syncrude and Suncor, Highway 63 branches out in three directions. Off to the west is CNRL, so massive that it has its own airstrip. Another branch juts straight north and leads to Fort Hills and Aurora. The silver buses wander up another branch that is called Conterra Road, passing another airstrip then Albian Sands, Jackpine, and Kearl Lake refineries. Finally there's a branch off Conterra Road that eventually reaches Firebag almost a two-hour bone-jarring bus ride from Fort McMurray.

Around those refineries, like leaves on those branches, sprout white tin construction camps. Buses feed the refineries from all those camps. There's Borealis, Millennium, Lakeside, Barge Landing, Albian Village, and on and on, all sounding like exotic resorts, not the sprawling white trailer villages they actually are.

All those refineries and all those construction camps suck up transport and finally disgorge thousands of workers with an efficiency and timing that puts the meat processing of an abattoir to shame.

For the first week of the shutdown, the crew had stood at the bus stop and stared silently at a passing yellow school bus filled with non-union workers. When sports teams play too many games against each other, insults get remembered, animosity builds, hurts get compounded. So it was with bus number 32.

Night after night, the crew—Pops, Stash, Dougdoug, and the rest—waited stone-faced at their bus stop as bus number 32 passed, always five minutes earlier than theirs. Oh, how they begrudged those minutes. That five minutes meant that the non-union workers were first to the camp, first to get the hot water in the showers, first at the dinner table, first to the telephones after the meal, first, first, first. Resentment filled the union members like the taste of pewter on their lips.

The coker just happened to be the first stop after the buses entered the front gates, and the last stop before the buses left the refinery. There was no use explaining to the crew that there wasn't some ulterior motive that determined their schedule; it was simply geography.

The feud had started with a skinny, pimply-faced little sucker who barked the loudest, caused the most fights, and never seemed to be around when the big boys went toe-to-toe. It was just a sneer the first day. But, like a tongue in a toothache, the crew couldn't stop searching him out as bus 32 rumbled past.

The next few days it was a giggle, a sneer, a one-fingered salute. The insults were reciprocated in kind.

Bus 32 was the enemy. Bus 32 was the lightning rod for all the real or imagined slights between the union and non-union workers. Bus 32 was going to learn a lesson in union solidarity.

After the foreman's warning, Pops called the crew together. That night, when bus 32 drove past, the entire crew turned their backs, just as Jason had ordered. Then, as one, they dropped their pants and mooned the bus. Pops was the only one who stayed upright, watching out for any company officials or cell-phone cameras on the bus. If he spotted any, he was to immediately call off the freezing of a dozen pink union buns.

Early next morning at the Toolbox Talk, the crew loudly complained to Jason about the obscenities inflicted upon their gentle union souls by bus number 32. No mention was made about the exposure of a dozen boilermaker buns to the cruel eyes of Bus 32 or the Fort McMurray winter. Too Tall, who had learned the welding trade during two years less a day in one of Canada's finest institutions, declared that he was shocked, yes, shocked and appalled.

A formal complaint was made. Pops the union steward made sure that the complaint was delivered late enough in the shift so that there would be no time for the accused to register a counter-complaint. Pops and the union crew could honestly say that they had warned the people on Bus 32. He just made the warning too late.

Pops made a special request to the Safety Nazi. "Stand with us at the bus stop," he said. "See what kind of dirt they inflict on us poor, innocent tradesmen."

At 5:15 sharp, Jason, the Safety Nazi, and several other foremen stood amongst those poor, aggrieved assault victims. Pops nodded to several of the workers as Bus 32 came into view. The crew aimed every cell phone camera they had scrounged that day at the passing vehicle.

As Bus 32 rumbled past the assembly, window after window on their side was filled not with grim faces scowling down at the union workers, but with the pressed hams of a dozen bald, pink, non-union rear ends.

"Isn't it beautiful when a plan comes together?" was Pops' only comment. His foreman suppressed a smirk.

Their plan didn't change the bus schedule one iota, but for days thereafter when it drove by, Bus 32 was empty.

(Email Day Fourteen)

From: Doug
To: Dad
Subject: Fort McMurray

Hi, Dad.

At lunchtime today, we discussed and planned the perfect heist. Modesty prevents me from telling you which worker was voted by the other welders as coming up with the next-to-perfect crime.

Last week, a young worker was killed on the highway between the refineries. His crew welded together a cross made of four-inch channel iron, took his hardhat, and drilled it onto the top of the cross. Then they took that cross and hammered it into the frozen ground with the blade of a front-end loader. It's about six feet into the ground on the side of the ditch where he was killed. That cross ain't ever coming out, not by human hands it's not. I counted five crosses in a four-mile stretch between refineries. It's a shame the government doesn't make enough money from the oilsands to construct a better highway.

A worker lost the tip of his finger when a pipe guillotined his hand. The report put the accident's cause down to a "blonde moment."

The time to leave Fort McMurray is when you have a favourite restaurant.

In one of the refineries in Fort McMurray, an X-ray company inadvertently exposed a group of welders to a dose of X-rays. It happens occasionally, and normally it's not a big deal. But it did happen, so an "incident" was declared and reprimand letters were sent out to the X-ray company. In their defence, the X-ray company stated that the exposed workers only received the equivalent of a head-to-toe X-ray.
One welder brought the incident up at the next union meeting and announced that he was going to seek counsel and was going to sue the X-ray company. Even though he was nowhere near the incident, he claimed that he was feeling the effects of radiation. When the union president asked him to describe his symptoms, the man stated that he noticed that his feet were glowing.

"It was so cold today, I got nine-inch nipples!"

There's a guy, "Ratboy," in the next construction trailer from us.

Ratboy went to a bar in Fort McMurray. At the bar he reached under the table and pinched the ass of the long-haired beauty in front of him. Longhair turned out to be a guy, who turned around and proceeded to punch out Ratboy.

Later that night, Ratboy picked up a girl and a case of beer and took her back to her apartment, where her husband quickly relieved Ratboy of the beer and then punched him out.

Hitchhiking back to the construction camp, Ratboy was picked up by a couple of locals who took him down a gravel road, relieved him of his money, and, oh yes, punched him out.

"You know, if I had to live it over...
I'd be living it over the liquor store!"

I'm getting tired Dad, real tired. I'll be glad to get home.

Take care,
Doug.

DAY FIFTEEN

(Jason by the Radio)

"Jason."
   CRACKLE.
"Jason Navotnick, come in."
   CRACKLE.
"Jason Navotnick, come in, please."
   CRACKLE.
"Come in, Jason."
   CRACKLE.
"Are you by the phone, Jason?"
   CRACKLE.
"Jason, are you by?"
   CRACKLE.
"Jason, are you by?"
   CRACKLE.
"Jason, are you by?"
   CRACKLE.
"Aww, Jason's not by, but he can be really, really friendly."
   CRACKLE.

(Double Scotch's Issues)

"I brought you in today to talk about these X-rays on your last batch of welds."

"What of them?" Double Scotch asked defensively. Jason, her foreman, knew that within this small, quiet woman beat the heart of a tigress, a tigress with a bad tooth. Looking into the face of death the foreman plunged in.

It had been Jason's experience that whenever he was faced with giving someone bad news, he wished to God he were giving it to a man. "Hey, buddy, you screwed up your weld test. If you screw up one more time, you will be fired. Deal with it."

For a foreman, giving bad news to a woman can have one of two consequences: either she'll cry or she'll wait until she gets back to her bunk. Then she'll cry. Either way, the foreman knew for the next couple of days he was going to feel like a puppy beater.

The only married couple in the crew were also welding partners. Scotch and his wife Double Scotch travelled the world welding for six months a year; for the rest of the year, they lay on some exotic beach. They had welded in Broken Hill, Australia, Kimberly, South Africa, and now Fort McMurray, Canada. They made their money in the First World and spent it in the Third World, where it would last the longest.

It was a rock star lifestyle that afforded them money, travel, and adventure. But you had to be very good welders or that lifestyle would come to an abrupt halt. Scotch and Double Scotch were good welders, very good welders—except now.

"Well, the test...," Jason began, trying to lighten the blow. "It's not baaad. But you were always the best. And these tests are borderline. You see here..." The foreman held up the X-ray of the weld. "There's slag inclusion on top of this horizontal weld here, and here. And right here on top, there's just a touch of lack of fusion. Not much, but it's... there."

Double Scotch's eyes flickered up at the X-rays, her cheeks

glowing.

"It's like you're having a hard time reaching the top of these welds. I've seen your welding before, and you're better than this."

Double Scotch's hands started to shake. Jason looked at her hands. He hoped to God she wouldn't cry.

"You'll just have to settle down and…"

Double Scotch jumped up, ripped off her leather welding coat, exposing her impossibly small denim shirt, and cupped her breasts. "It's because of these!"

Jason didn't know whether to shit or wind his watch. "Pardon me?"

"It's because of these! These!" She continued to cup her breasts, aiming them at her foreman, making her points.

"Gwen!"

"I told him I didn't want them! I told him they were…"

"Gwennnnn!"

"I told him they would get in the way. But oh no! He said it was like running his hand up the wall to turn off a light switch, and I said…"

"For Chrissakes! Gwen!"

Gwen skidded into the doorway. Double Scotch continued her rant, all the while cupping her breasts. Jason, half-sitting, half-crouching, mostly cowering behind his desk, threw up his hands imploringly between Double Scotch and Gwen.

"I can't get in there close enough! I have to keep my elbows out instead of tucked in. I have to work around them all the time. My arms get tired!"

"Implants?" Gwen said.

"Like two traffic cones!" Double Scotch nearly shouted.

Jason opened his mouth, decided against it, and sat, his head against the wall.

"And when I can get in close, I burn them!"

"Okay," Jason said. "We'll take you off the delicate stuff and give you some pad welding until you, ah, adjust."

"I'll never adjust! I'll never adjust! That bastard may like them, but he'll never get to touch them!" Double Scotch exited, slamming the trailer door behind her.

The room was heavy with the tiny woman's bitterness. The foreman and his secretary stared at each other. Then she sniffed, and walked back to her desk. The foreman blinked at the empty doorway. Gwen's voice echoed into the large outer office:

"All men are pigs."

DAY SIXTEEN

(Lobotomy's Final Phone In)

"Golden and Fliese, Gwen Medea speaking."
"Hi, Gwen."
"Hi, Lobotomy."
"I'm sick."
"You know, Lobotomy, Tim, you can't just keep on doing this, phoning in sick every second day."
"But I'm really, really sick."
"Well, how sick are you?"
"I'm fucking my sister, how sick is that?"
"Tim, that wasn't funny in the seventies and it's not funny now. You come into work now!"
"Oh."
"Now, Timmy."
"Okay."
"Now."
"Alright, alright, I'll be there. Besides, I don't have a sister."
"Now."
"You know, Gwen, you sure are sexy when you're mad."
"Lobotomy! If you're not on that bus in fifteen minutes, I'll knock out your one good tooth and all that will be left of you will be cocaine and hooker spit!"

Lobotomy gasped.
Gwen slammed the receiver down.
Then giggled.

DAY SEVENTEEN

(Lunch Break)

"I was watching this iron worker try to pick up an eight-hundred-pound piece of scrap."

Pops paused to open an aluminum lunchbox covered in stickers from several unions and various brands of bananas. What small part of the lunchbox that wasn't covered in logos was covered in grime. His hands dug into the box and extracted an egg salad sandwich. The humid egg-and-onion smell wafted over the two hunched men.

"Did he do it?" Stash asked.

"Well, he got his fingers under it."

"Too bad. If he crushed his fingers, his nose would never be the same."

The grizzled man stopped for a moment and focused on Stash. A small smile broke through his dusting of day-old white beard.

"So," he continued, "as I'm watching him trying to lift this thing, the ironworker foreman is watching too. So I figured I'd hang back and see what happens. After the foreman watches him wrestle for a bit, he walks over to the ironhead and says to the iron worker, 'You can't solve all your problems with brawn. One day, you'll have to break down and use some brains. But I don't think that's possible.' Then the foreman turns and walks away."

Stash grunted.

"So I wait until the foreman is gone and I walk over to the guy and say, 'I think your boss just insulted you.'"

"What'd the iron worker say?"

"Nothing at first. Then he looked where the foreman had gone and said, 'Yeah.' But he drew it out like 'Yeeeaaaahh,' like he had just discovered electricity."

"Never mind," said Stash. "I was working up north with a guy, and I asked him what time it was. He looked at his watch and said, 'We're gettin' there, we're gettin' there.' Then he walked away."

The two men munched sandwiches made from the construction camp kitchen. The choice of bread today was white with bland egg or white with bland ham, and a pickle. Spice was used sparingly, but refined sugar was plentiful. The construction workers could always tell who had been in camp the longest because the worker was probably fat. Fat, like no-longer-able-to-see-important-parts-of-his-body fat.

"What've they got you doing?" asked Pops.

"Right at the top of the coker."

"How's the view?"

"You can see McMurray."

"You know, the first time I climbed one hundred and ninety feet, I thought somebody was squeezing my nuts. I sweated right through my gloves. My mouth was dry, but every other part was soaked."

"You piss yourself often?"

"Every time I climb."

Stash poured himself a coffee from a worn green metal Thermos bottle, the kind of instrument that gets splashed down every week or so with the understanding that hot coffee acts as its own sterilizer.

"How's Baker?" he asked.

"He says he can feel the backs of his hands. Below that, it's all numb, like going to the dentist."

"Is he off the ventilator?"

"Yeah. They had his head in that hula-hoop deal with screws going into his skull to hold his neck steady."

"I hear Ralph is taking it hard."

"Yeah. He was off most of the summer. Didn't leave his house once."

The men munched in silence.

"See that guy over there? No, the guy in blue."

"What about him?"

"That's the guy Jason talked about at the Toolbox Talk."

"What did he do?"

"Tried to line up a twenty-foot steel beam with his finger."

Both men grunted.

"You tried the carrots? Kinda tasteless."

Pops, ignoring Stash's carrots, started in on another story. "Down home, there was a guy at a welding shop that was into the casinos for $100,000. The shop he worked at gave the workers $40,000 for each lost finger. I guess the guy figured, three fingers he's outta debt and $20,000 ante for the next game."

"I know where this is going."

"So one morning, he walks right up to the metal shear and sticks three fingers under the knife and pulls the lever."

"What happened?"

"He cut his bloody fingers off! What do you think happened?"

"I know, but what happened after that?"

"The foreman runs over to the shear, picks up the three fingers in some toilet paper, and takes them and the guy to the hospital."

"You gonna finish those pickles?" asked Stash.

"No, here, take them."

"So what happened?"

"So they sewed his fingers back on, fired the guy, didn't pay him any money, and his fingers are almost useless. These pickles are really crunchy. Oh, and one of the sharks he borrowed money from? Said that if he didn't pay up, they'd break his legs."

"Speaking of legs, trade you this drumstick for some of that ham."

Both men munched in quiet contemplation.

Pops flicked a piece of ham fat from his cheeks, his mind in deep thought. "You don't ever want to have a heart attack in a pig pen," he said.

"What brought this up? The ham?"

"No, I mean it. If you ever faint in a pigpen, at first the pigs will nudge you with their snouts. If there's no reaction, they'll give you a nip. If there's still no reaction and you don't fight back, the next bite will take a hunk out of you. Once one does it, they all join in. You don't ever want to have a heart attack in a pigpen."

"I'll try not to," said Stash.

"You know, my ex thought we are all pigs. You try to have a nice interesting conversation like this, like you an' me are having, and she would get all upset and just stomp out."

"My wife too."

"Yeah. They just don't understand."

(Email Day Seventeen)

To: Dad
From: Doug
Subject: Ft. Mac

Hi.

There's an ironworker here who walks with a limp. What happened was one night a group of riggers in a hotel room were trying to rid the world of alcohol and they got on the subject of how good they were at climbing steel. Each one loudly bested his neighbour in the high rigging contest. Our hero protested that he, and only he, had more strength, more agility, and more stamina than any other construction worker in that hotel room.

"Prove it!" was the shout.

So our hero stripped the sheets from the bed, tied them together and then to the bedpost, and flung the makeshift rope out the window to the pavement three stories below.

"I'll go down this rope, touch the pavement, and climb back up in under ten minutes!"

Then he went out the window.

All would have gone as planned until the crowd that was sitting on the bed jumped off and ran to the window to watch. The bed, released of its weight, shot to the window. The jerk snapped the makeshift rope. The rigger went down the three stories with the rope still in his hands.

He was correct on one point though. He did touch the pavement.

Doug

DAY EIGHTEEN

(Gas Monitor)

"Gather round, everybody."

The crew stood around Acastus outside the trailer.

"Okay, we're handing out your personal H2S monitor. We've had a couple of incidents and until that's resolved, everybody wears one all the time. The only exceptions are the welders who must take it off while they are welding.

"Clip the monitor to your breast pocket close to where you breathe. It does you no good inside a pocket or in the lunch shack.

"H2S, hydrogen sulfide, or sour gas, is that rotten-egg smell that occurs naturally in swamps, garbage dumps, and in our case, oil refineries. It's the natural by-product of rotting organic material.

"These monitors are set to start beeping at five parts per million of sour gas. After five parts per million and your monitor starts beeping, you are to immediately clear the area and head to the muster point. Believe me, after a good whiff of sour gas, nobody will ever have to tell you twice to get your ass away.

"Hand them in at the end of shift and if they have been activated for any reason, tell us so that the monitor can be re-calibrated..."

BEEPBEEPBEEPBEEP!

Twenty sets of eyes turned to the tiny woman in their midst. Double Scotch smiled a red-cheeked smile.

"'Scuse me."

(A Lesson in Love)

"Today, Sonny Jim, I will make you a man!"

My apprentice Dougdoug looked at me. "Does this entail whips, chains, and midgets?"

"Not yet, but hold that thought. No, today I teach you the fine art of... of... wait for it... steel fabrication!"

"Oh... joy," Dougdoug shrugged.

"Look on the bright side."

"There's a bright side?"

"Well, kid, when you work with me, you get to say the word 'fuck' a lot. Try that working at Wally World."

"Bonus."

"I can see you are underwhelmed, my young appren-tye."

"Well," Dougdoug shrugged.

"I'm not kidding, kiddo. We have a tray to replace inside the coker. It's a pretty good job, really."

"I'll take your word for it," Dougdoug said.

"Okay, kiddo, write this down. We need a grinder, an extension cord, extra grinding wheels, cutting torches, hoses, gauges, a welding machine and welding cables, a couple of quick slaps, welding rods, mark-out chalk, and don't forget the cuddle. Now read that back to me."

In the middle of the reading, Dougdoug stopped. "What's a couple of quick slaps?"

I slapped Dougdoug's face, twice.

"Hey!"

"So now you get the cuddle."

As I opened my arms wide, Dougdoug ducked.

"Very funny, Rance," he groused.

"It's construction, Dougdoug. Don't trust anybody. Just be glad I didn't send you to the tool crib for a four-inch Fallopian Tube."

"A what?"

"Eight inches are better, but a four-inch Fallopian Tube will do."

Dougdoug shook his head as I cackled.

I kept up a steady stream of banter as we assembled the equipment. Once everything was in place, the lines laid, the string of safety lights hung, and the hand tools assembled, I called a break. We sat inside the coker, swinging our legs over the edge of the tray, three stories above the next level.

"You know, kiddo, I love this... this... making shit. The grinding, the cutting, the making something out of nothing. I love it."

Dougdoug gave me a quizzical look.

"If I have to explain it to you, then you don't get it. All those 'Human Resources' experts, with their white shirts and red suspenders, they all talk about job satisfaction, but if you want real job satisfaction, try working with your hands. I love it. An' working with metal is as old as time."

"What do you mean?" Dougdoug asked.

"Working with metal is not about history; working with steel is history. You ever go to Sunday school?"

"Yeah."

"You know Abraham, the father of Isaac and the twelve tribes of Israel?"

"Yeah."

"Well, in the Bible, he was from Ur of the Chaldees. You know what they did at Ur of the Chaldees?"

"No."

"They mined copper. If you go to Iraq today, you will still find those same pits at Ur, pits that they dug out of the sides of the hills to get the copper. They dug with their bare hands, three thousand years ago."

"Humm." Dougdoug started to sound interested.

"They would pound out the copper mines using rocks, their bare hands. Yeah, and their Personal Protective Equipment was a loincloth.

Once they got enough ore, they piled the gravel and copper in fire pits, and then they would burn the rocks. For days."

"Pretty tough apprenticeship," said Dougdoug.

"Wait, it gets better. Once they had that fire blazing, they all sat around it, heads stuck in that smoke, with nothing on but their sweaty loincloths and sweatbands holding long pipes made of leather or clay. They blew into those pipes into the base of the fire. Breathing in copper fumes gives you chills and convulsions, a metallic taste in your mouth, vomiting, and a whole lot of other crap."

"Ouch."

"Not only that, after a couple of months of inhaling copper fumes, it turns your shit blue. That's when you know you've got a real good dose."

"Blue shit?"

"With streaks of blood in it. Read that in a Safety Book."

"Nice."

"So they'd sit there for days, their heads in smoke, blowing into those pipes, while someone stoked the fire, refining that copper. That's where they got the expression 'Blow, don't suck.'"

Doug grimaced.

"After a couple of weeks of blowjobs, they would dig the bloom out of the fire. A bloom is just a bagel-sized lump of copper, twigs, rock, dirt, and crud. So then they would all sit around and beat the shit out of that bloom. Finally, after about a month of beating their lump, they would get enough copper to make one sword. One bloody sword. Abraham must have been a metals trader rather than a metal worker."

"How do you know that, Rance?"

"Because he gets on his camels and heads off to so some trading in Canaan. If he actually made the stuff, he'd be dead long before he had twelve sons. And you know why Abraham had twelve tribes?"

"Why?"

"TV was the shits."

Dougdoug shook his head.

"Well, let's get to it," I said, reaching for a welding electrode.

"Hmmm. I see my working-class humour is lost on you."

I explained to Dougdoug how, in laying out metal, there are no absolutes. Sometimes you use the measuring tape, sometimes you make a template, and sometimes you estimate. I told Dougdoug how to use his eye to line the metal pieces up, how to use his bare hands to feel for the "high-lows" or distortions, how to use his fingers to feel for the smoothness of the finished joint.

"So, Dougdoug, do you know what part of your body you use most when you're screwing?"

"Huh?"

"Do you know what part of the body you use most of when you are making love?"

"That's a stupid question."

"It's a trick question, Dougie."

"It's..."

"Wrong." I barked. I wiggled my fingers in front of Dougdoug's face. "It's your hands."

"Oh. You meant with other people." Dougdoug grinned.

"Very sharp, kiddo. Mind you don't cut yourself. You use your hands to feel, touch, probe, just like you would working on your old lady. I mean, for fine tuning, use your hands, use your hands."

"I wonder how my girlfriend will like it when I start marking her out with a chalk-line."

"Let alone the grinder," I chuckled.

The day passed. The coker's walls echoed with my old man's cackle, rising as the punchline of each of my wonderful stories came to an end. Our laughter, echoing inside that giant vessel, stopped every so often as we ground the metal, and finally to listen to the ripping-paper sound of an arc welder.

I told Dougdoug how a welding rod works the way the force of water on a hull of a ship pushes the ocean aside. The molten arc pushes the metal outward and aside. But when the arc passes and the metal cools, the steel behind the arc is pulled together, only now it pulls back more than the metal was originally.

"This is all natural, Doug. You can make the forces of nature work for you, and make it look easy, or you can beat the hell out of the steel and make it look like a golf ball."

"You really like doing this, don't you, old man?"

"Working with steel has been my only absolute. It's the only thing that I can depend on. I instinctively know how steel will move, how it distorts. I just know. Try saying that about a woman," I added with a half-smile. "You know what I think, Sonny Jim? I think... every major change of history or development of mankind has been to do with metal."

"The pill?" Dougdoug asked.

"I said mankind."

Dougdoug chuckled.

"I always found women to be... hard to... Well, I dunno, if a guy doesn't like you, he punches you. I can live with that. I think, if there is love, or call it what you want, the best construction workers like us can hope for is loving your work, like this." I pointed to the work we had just completed.

I got up and turned away from the young man, making like I was examining the weld. I was too embarrassed to open my heart and say this to another construction worker face to face. I hoped he didn't hear my old man's voice quaver.

"Making things with your hands is... is the best job in the world."

DAY NINETEEN

(You Can't Drink Him Pretty)

Pops and I stood high up on the coker away from the crew, looking out over the silver and tar towers of the refinery. When our eyes did meet, it was quick.

Pops did the talking. He needed to talk. But he needed me to be there.

"He's gone... Lobotomy is. The kid never stood a chance. I knew his mother. She was absolutely beautiful. Crazier than a shithouse rat, but beautiful, once. Lobotomy told me the last time he saw her, she was pushing a shopping cart full of clothes down a Regina back alley. His own mother, for Pete's sake.

"He can be so... so innocent and batshit crazy at the same time.

"Some people look for trouble, Lobotomy didn't. He was like those Play-Doh molds kids get at Christmas. He'd just takes on the personality of whatever person he attached himself too. I told him once that the penitentiary is full of guys that just went along.

"I had him as my apprentice in Fort Saskatchewan. Best apprentice I ever had, for a while. Then it dawned on me that Lobotomy couldn't make his own decisions. Shit, he couldn't even live his own life.

"He started to follow me around, sat when I sat, picked up the tools when I did. When I noticed him starting to wear the same type workclothes I did, that really started to creep me out. I made some noise about it being time for him to work with other tradesmen, see different ways to work. Really? I just wanted to get away from him.

"The morning I told him that he was on the other crew, Christ, it was like kicking a puppy.

"You should see his tattoos; they're something else. He keeps 'em hidden, has to. Got involved with a gang of skinheads, couldn't win a fight unless there was six of them up against a nun in a wheelchair, but they were pretty good with the tats. So Lobotomy got all tatted up. Really ugly. Other than paying for it, with all that blue crap all over his skin the only way he was ever going to get naked with any sane woman was if they were in the same mass grave.

"So last night he goes into McMurray and buys it.

"Him and his lady of the hour go upstairs in the hotel. Right in the middle of them bumping uglies, he made the mistake of ripping off his T-shirt. He's got this huge swastika on his chest, and this Aryan Nation Adolf-fucken-Hitler crap all over that skinny little body of his.

"The hooker, she looked up at all those tats and screams. Lobotomy said she didn't so much scream as gag. He said that a couple of times. He said her lips curled, and she wrung her hands, like his body was covered in green snot.

"The hooker grabs her clothes and runs out of the room, screaming. She's running down the hall, trying to put her clothes on, and all the time crying.

"Lobotomy's trying to get his clothes on and go after his money. I think he really wanted to calm her down. Make her not get sick of the sight of him.

"She runs into the bar, still putting her clothes on, and she runs over to some ironworkers and asks if she could sit at their table. I hear she was pretty much babbling.

"She's sitting there... hunched over, really... when Lobotomy comes running in. She screams, 'That's him!'

"The guys in the bar put two and two together and get five. They start calling him a pervert and a freak and start laying the boots to him. The whole bar joined in. About five guys chase him out of the bar into the parking lot.

"He made it to his truck and took off with all these guys hanging on. By this time his eyes were so wide he just wanted to kill something, anything. He said he couldn't get that woman's retching out of his mind. Really? I think he wanted to kill himself.

"Across the parking lot, he thinks he sees a railway fuel-tanker car parked. So he aimed the truck at the car, and floored it. The bar's security guard says by the time the truck reached the end of the parking lot, it was almost airborne.

"I guess he planned to take out that fuel train, everybody in the bar, and himself in one gigantic fireball.

"Everybody in the lot dove for cover.

"But what he thought was just a piece of grass between him and that tanker car was a ravine, some pine trees, a creek, four or five boulders, and a chainlink fence.

"You go there today and there's just a long line of truck parts starting from the parking lot into the ravine and up the other side. The only thing that actually bounced off the tanker car was his side mirror. Besides, the thing turned out to be a water tanker.

"He's laying in the dirt all broke up when the crowd got to him.

"Lobotomy's gone."

DAY TWENTY

(Stretches)

"Ricky! You're leading the morning stretches today."

"Aw, I don't want to. And don't call me Ricky."

"Tough shit. Lead the stretches."

"You can't make me."

"Can too."

"Can not."

"Can too."

"Can not."

"Too."

"Not."

"Rance!"

"Aw. Why don't you get Bob? He's never led the stretches."

"Because I told you."

"Yeah, sure. You don't get Bob to lead the stretches because he's connected. President of just one bike gang and you get a free ride."

"Lead the stretches."

"Okay, okay. I'll lead the stretches."

"Listen up! Today, Ricky is going to lead the stretches."

"Okay, everybody!... Gee, thanks, I'm underwhelmed by your enthusiasm. And don't call me Ricky. Anyway. Everybody watch me!

"Put your left hand in!

"Put your left hand out!

"Put your left hand in!

"And shake it all ab—"

"Next! Bob! You lead the stretches. Ranson! I got a job just for you."

(Terminated)

"You must be joking." Jason glared at the typed paper that Acastus slapped on his desk.

"Safety is no joke," the Safety Nazi growled.

"If she goes, her husband will too."

Acastus shrugged.

"I've already lost Lobotomy. With both her and him gone, that..." He pointed at the typed paper. "...shuts me down."

Acastus shrugged again.

"That's my welders, all except Pops! I won't be able to finish...."

"That's your problem," Acastus replied coldly. "She's terminated."

"For this bullshit charge?" Jason threw the paper toward the wastepaper basket. The single-spaced typed paper teetered on the edge of his desk. The Safety Nazi glowered at this show of open disrespect.

"I warned her."

Jason glared at Acastus. The foreman's worn and battered hardhat was covered in years of company logos, union stickers, and gouges, all acquired from dozens of projects across all of Canada and a couple of American states. He had two more just like it tumbling around in the back of his truck. The Safety Nazi's hardhat gleamed white, virginal.

Jason stood and looked down at Acastus. Sitting at her desk with her back to the foreman's door, Gwen tensed.

"No. I'm not going to do it," Jason said. "Wearing her goggles incorrectly is a bullshit charge that will stop this job cold. No. No fucking way."

The Safety Nazi's eyes gleamed. "Put that in writing."

"Look. You watch the guys come out of the coker. They're all wearing these goggles at the end of their noses. Either that or they're hooking them onto their hardhats, or they're ripping the rubber out to let air in. These things fog up! You can't see out of them. They're dangerous!"

Jason threw his goggles on the report. The mono-goggles and Acastus's report teetered on the edge of the desk. They fell onto the floor, together.

"It's a fact that they reduce eye injuries."

"It's a fact that if you report an eye injury, you are in deep shit! So nobody reports it!"

"Well, I'm reporting this. I didn't make the rules."

"You stand at the end of the shift and you look at the guys' faces. Everybody's cheeks are covered in dirt and you know why?"

Acastus blinked.

"Because they're taking their glasses off and on all day with dirty gloves!"

Acastus shrugged and turned to leave.

"Somebody's got their career tied up with these glasses and they won't admit they screwed up!"

Acastus continued to retreat. When he reached the safety of the open door, he spun towards the larger man. "She's terminated, with or without you!"

The trailer reverberated in silence after the door slammed.

Gwen stood in the office door, the winter sun glowing behind her, making her neck and shoulders all fuzzy. Jason was surprised at how delicate she appeared.

"What are you going to do?" she asked.

"The most I'm going to do is to give the entire crew a verbal. That'll be the end of it."

Neither of them believed him.

Suddenly, there was a noise like a shotgun blast at close range as someone kicked in the trailer door, followed by the sound of a chair hitting a wall followed by the sound of a newly castrated bull. Scotch had arrived.

"Fired! Fired! You fired her?"

Several of the crew tumbled in after the large man.

"No," Jason said. "I didn't fire her!"

"Well, you're gonna!"

"No, I won't."

"That's not what Safety said!"

"The Nazi's an idiot...."

Scotch bolted around his desk at his foreman. He was blocked by several hands. The noise level increased as desks were scrapped, chairs overturned, and men shouted as Scotch and Jason tried to get at each other. Several more crew members joined in. Soon they couldn't even move, let alone fight.

An airhorn blasted within six inches of the struggling men's ears. Everyone winced and froze. Gwen Medea, all ninety pounds of her, stood amongst the fighting mob with her tiny hand holding a smoking airhorn.

"Sit down," Gwen told the children.

The men looked at her.

"Please."

One after another, the struggling mass slowly disentangled themselves and sat with their hands between their knees like a chastened Grade Three class. Finally, Gwen Medea was the only one standing.

"Jason, the Project Manager is on line one, the Safety Coordinator is on line two, and the union is on line three. I sent for the Job Steward." With a dismissive sniff to the seated men, the tiny woman returned to her computer. Every man in the room stared at the empty door.

The crew leaned in and listened to Jason's side of each telephone conversation. Scotch's face softened a tiny degree when he heard Jason speak into the phone.

"He continually crosses the Safety Silly Line! He's the one getting in the way of safety. He's the enemy. He even comes close to walking through the site and everybody stops working. They just stare at him. She was wearing them! Just not the way he wanted."

Pops arrived and glared around the room. Several crew members got up and left. Double Scotch, eyes red, arrived and took a seat close to her husband. He squeezed her tiny hand. At her appearance, Scotch's eyes turned murderous. Pops noticed a pink scar across Scotch's nose that he'd never seen before.

"Where's the little weasel... right now?" Scotch asked the room. Double Scotch's hands reached out at his words, grabbing her husband's arm.

"Off-site," Pops lied. The men nodded.

At one point, Jason covered the speaker and addressed the room. "What's the crew doing?"

"Working," Pops lied. This time nobody believed him.

Right then, the crew were standing in small groups, or leaning on scaffolding rails, or bunched over the counter of the tool crib, all discussing the same thing. If the loss of production dollars in salaries, stalled cranes, idle loaders, forklift rentals, trailer leases could be magically calculated, the Safety Nazi's display of bullying a tiny woman was costing the company thousands and thousands of dollars every minute.

The job froze, waiting.

Jason wiped sweat from his hand and telephone receiver. The telephone conversations with several layers of bureaucracy dragged on. All the conversations started to sound the same. One by one, the crew got up, stretched, and left the office to the familiarity of the coker. As the afternoon sun set through the silver towers of the refinery, only four people remained in the office: Scotch, Double Scotch, Pops the job steward, and Jason the foreman.

The final phone call over, Jason leaned back in his chair. In a tired voice, he spoke to Double Scotch,

"This is a bullshit charge, but consider yourself verbally warned."

Double Scotch, forty, with welding experience all over the industrialized world, looked like a little girl spanked. Scotch bristled. Jason held up his hand.

"I know, I know."

"Well?" Scotch asked.

"Go back to work," Jason said, his voice hoarse.

Pops looked meaningfully at his watch. Taking the clue, Jason said, "Get on the bus. Let's call it a day."

Jason watched her hand as Double Scotch put it on her husband's arm. Before Pops ushered them out, Jason said, more to himself than anyone lese, "Safety's got the company convinced that they got God on their side. Big fucking oil company, scared shitless of safety."

"*Gott mit uns*," Pops said.

Jason nodded.

Pops turned to Double Scotch. "It means 'God is with us.' It's what the Nazi SS had on their belt buckles."

An hour later, when Gwen went to turn out the lights in the foreman's office, she found Jason Navotnick staring across the office at the blank white wall.

Jason turned to Gwen. "It's one thing to fire somebody," he said. "It's another to like doing it."

DAY TWENTY-ONE

(Too Tall Won't Be With Us Anymore)

"Too Tall won't be with us anymore."

"What happened?"

"He and his millwright buddies went into McMurray last night and must have really got drunk. After the bar and a fight, they stopped at a gas station. While the attendant was busy, Too Tall and his friends stole the gas station's flag. They got them on tape doing it."

"That's a pretty big flag."

"Yeah. I guess all three of them had to sit in the front seat of Too Tall's Honda because the flag took up the entire back seat. So after they stole the flag, they tried to hide it in the millwright's apartment. It took all three guys to drag the flag up the three flights of stairs. Woke everybody up in the apartment block with all the bumping and giggling. The flag was so big that even all balled up it still took up half the guy's bedroom.

"And filthy. Aw gawd, everything the flag touched is covered in oily dirt, the car, the apartment block's hallways, the apartment, them, everything. Finally they got pissed off and threw the flag off the balcony. When the cops finally showed up, you didn't have to be no Sherlock Holmes to figure out who stole the flag.

"Too Tall won't be with us anymore."

DAY TWENTY-TWO

("Got Ten Bucks?")

"Got ten bucks?"

Dougdoug looked at Stash's outstretched hand.

"Ten bucks," Stash repeated, flicking his fingers. "Cheque pool."

"Gambling?"

"No shit, Sherlock! Gimme ten bucks."

"You don't have to, Dougdoug," I interjected. "Stash is just a frustrated encyclopedia salesman."

"Well, how much can I win?"

"So far," Stash said, "eight hundred and change. But I haven't hit the electricians. Could get to be over a grand."

Dougdoug, juggling his coffee cup, reached into his pocket and pulled out a ten-dollar bill.

Pocketing the ten, Stash fanned a deck of poker cards in front of the young man. "Pick one, Slick," he said. "Write your name on one half. Then rip the card in two and give it back to me. No! No! The half with your name on it. Sheesh... newbie."

Stash crabbed off down the construction site trailer, eyes flicking left and right, searching men hunched over their meals.

"Every time I see Stash in action, I hear rattlesnakes," I said. "He feels like if he beats you out of some money, he beats the system. It's nothing personal—you're just standing between Stash and his money."

The young man leaned towards me as I continued.

"Keep the other side of the card. It's your receipt. It's always good to have proof that you actually did pay for the card. Besides, Stash is running the cheque pool."

"Oh?" Dougdoug flipped the half-card over, studying it.

"Yeah. Normally the guy running the pool keeps ten percent. But nobody ever knows for sure." I shrugged my shoulders. "Every time Stash has a scheme or a project, somebody gets hurt. Nobody's ever called him on it, nobody."

"I wonder if I can get my ten back." Dougdoug fanned the card.

"You're better off with the bigger cheque pools for being on the up and up. The ones the stewards sell. You can tell which ones they are because they got the cards already made up. Not this Mickey Mouse playing card shit."

"Gee, maybe I really should ask for my money back."

"Forget it. Who knows, you just might win. But the big ones are the official cheque pools."

"How big?"

"Last big one was a hundred thirty thousand."

"Wow. Every payday?"

"No, not usually, but on the long weekends or holidays, yeah, a hundred and thirty thousand isn't out of the question. Back in the day, they would make the draw for the big one, and then hand the winner two garbage bags full of tens and twenties. Then the poor bastard would have to run a gauntlet of these whack-jobs to his car, drive back down the highway to Edmonton with all that cash in his trunk, and then try to put fifty grand in small bills into the bank without attracting any attention. The cops caught on real fast.

"Now when you win, about four of the largest sonsabitches you ever seen in your life walk up to you. They hand you a cheque. After they slap you on your back, they escort you to your room at the camp, watch you pack your bags, walk you and your gear out to your truck, and if need be, they'll stay with you all the way to your bank in

McMurray. And they stand there and watch you deposit the cheque in the ATM."

"Wow."

"The big ones pay all the taxes and donate a portion of the money collected to charity."

"A hundred thousand. Holy smoke."

"Well," I said.

"Well what?"

"Usually after about $60,000, they start dividing the pot in two. Last September long, a double pot was won by some guy in the office, but the second pot was won by a university kid who had just quit and was packing up to go home. The stewards caught up to him in the camp parking lot. Can you imagine? You're all packed up, going home to university in your truck, in the parking lot, and four of the biggest bastards you ever seen are running after you trying to flag you down. And all of them are smiling."

"We all felt good about that."

"What about poker?" Dougdoug said.

"They ain't games of chance. Not around McMurray. Those games don't have a damned thing to do with luck. I've seen a poker game that had so much money in the pot, you couldn't see the table. A guy lost everything in that game. We didn't see him for three days, just lay in his bunk, staring at the ceiling."

"How much?"

"They say he worked that year for free."

"I'm pretty good at cards."

"I thought I was pretty good too. One time, me and a couple of guys played this shark. We didn't have fifty cents between us, but we wanted to play this... this barracuda, just to say we did. We were joking and laughing, but halfway through the game, I realized that each time the pot was won, the next hand went to the guy on the right. To this day, I don't know how he did it, but he controlled the game so well, every hand he'd move the pot one place to the right. That son of a bitch moved the pot wherever he wanted to. That sobered me right up. Those games don't have a thing to do with luck."

"Well..."

"And you don't know who you're playing against. Back in the day, one of the camps started stinking. They thought it was a dead raccoon that had crawled under the camp and died. So they dug around in the snow, and found it was some guy wrapped up in plastic garbage bags shoved under the trailer. With the number of men coming and going through here, there's not a hope in hell of his killer ever being caught."

"No leads? Nothing?"

"Never heard. So do yourself a favour. If you are going to play, go down to the casino in McMurray or leave the poker playing at home with Granny."

Dougdoug flipped the half-card in his hand, studying it.

Men were stretching and moving in their seats in anticipation of going back out into the cold.

"Once in a while, the guys try to change the game," I continued. "So, just before the cheques are handed out, they all put their money in a hat. Once the cheques are handed out, the guys gather together and they read the last four serial numbers on the cheque, and then the cents column in the amount. Those six numbers gets you a poker hand. Best poker hand wins."

"Seems complicated," Dougdoug said.

"And the game's not foolproof. The person making up the cheques can control the game."

"They wouldn't."

"Never say never. When there's thousands of dollars on the line, never say never."

"Maybe I'll keep my money in my pocket."

"Too late, you've already bought in."

"I have not!"

"You're here, aren't you? Stash, or someone like Stash, will be back again and again. And you know what? They're like alcoholics. They get mad if you don't drink with them, or gamble with them."

"G.A.?"

"Gamblers Anonymous. Stay in the trade long enough and you'll have more than a passing acquaintance with either A.A. or G.A."

"Is everybody around here nuts?" Dougdoug said. "I think I'm the only sane one in this whole crew."

"You cut me deep, kid."

"Okay maybe you're goofy, but in a nice way." Dougdoug smiled.

The new apprentice fingered the card. The quiet was broken by a great shuffling of feet. The crew had stayed in the lunch trailer far enough past the end of the noon break to show the bosses they were independent, but not long enough to get yelled at.

I looked at Dougdoug. "I been around a long time and I do know this: if there's going to be a fight, it'll be because of money, booze, or women."

"So?" Dougdoug asked.

"The only two women we got on this crew scare the crap out of everybody. We're in camp now, so booze is out. That just leaves one thing: money."

I looked at Dougdoug.

Dougdoug looked at the card, and said:

"Wanna buy a card, Rance?"

DAY TWENTY-THREE

(Spider)

On a winter's night, the steam clouds from a hundred vents hang silent above Albian Sands Extraction like a formation of dirigibles. The lights make the bellies of those steam clouds glow a mottled orange. When there is no wind, those silent clouds spread, join, then slowly sink between tanks and passageways, making the edges of the buildings and pipes glow a soft orange. You half-expect to stumble over a body.

The Extraction Building is where boulders and sand go one way and crude oil starts going the other. It's an ominous, dripping place that shakes and smells of tar and damp and a touch of fear. The massive conveyors growl and hurl acres of rock and sand and boulders into giant vats. The tumbling, Toyota-sized boulders growl and foundation-shaking thuds rumble the building when the sand gets ripped from its treasure. On the death-cold cement floor, pools of water tremble in cadence with the thump of powerful motors. When they walk past, workers give those tanks space.

The building has the smell and feel of newly tarred roof beside a cold mountain waterfall. Circles of shimmering gold lights bleed down the walls in futile shafts of light that give the walls dimension but no illumination, so the roof of that huge building is lost in an indistinct black.

Me and my apprentice walked through the mist, sidestepping a six-ton front-end loader that "meep-meep-meep"ed past us. The yellow-and-rust steel dinosaur pushed the fog aside, and just as quickly disappeared, leaving the smell of burnt diesel in its swirling wake.

"This place reminds me of Mordor," Dougdoug said.

"Huh?" I grunted.

"*Lord of the Rings*. You know, Mordor."

"You mean with all this yellow fog and the trucks an' the scary noises? Why would you say that?"

Dougdoug smiled.

"Guess what that makes us," I said.

"Orcs?"

"I prefer hobbits." I snorted. "You know, laddie, in the summertime when the doors are open an' there's no fog, this place looks like one of those huge buildings they got down in Cape Kennedy, the ones with the huge doors where they put together those rockets. I saw one once on one of my trips to the Keys."

The apprentice stopped mid-stride. He knelt and picked up something, and held it in his gloved palm. Both of us studied the dot. It was a spider. There was nothing special about it. The spider was just an ordinary black spider. This one had a slight limp, and rather than move around and explore, the insect sat in Dougdoug's damp glove. Eight eyes stared back at us.

"Musta rode in on the conveyor." Dougdoug studied the insect.

"Quite a ride." Both of us listened to the rumbling of a particularly huge boulder tumbling in that vat behind the mist.

"First bug this year," I said. "It's still February."

Dougie was quiet for second. Then he looked up. "Can you imagine? You're sittin' in your spider-hole, all fat, dumb, and happy, waiting for spring, and this gigantic claw comes down and grabs your whole world and dumps you into a truck, and then dumps you onto a conveyor, and then dumps you into a bin..."

"I know, ruins your day. C'mon, kid, let's go."

I stopped, then looked at the young man as Dougdoug continued to study the insect. Dougdoug spoke.

"Gee, it makes you think."

"About what?"

"About just how many other animals get eaten by these machines. After we're done with it, that sand comes out as white as talcum powder. After we're finished with it, the ground's just sterile... dead."

"It's a spider."

"You can't even grow weeds on the shit after we're done with it. Where'd all the animals go? You can't hide from something that takes a twenty-foot bite. Nothing burrows that deep."

"An' they boil the soil to get the oil. C'mon, let's go, Mr. Save-The-Planet."

"You know in that movie when that Death Star blows up that planet?"

I put my hands on my hips. "Where are you going with this?"

"Well, that's what it must be like to these insects. Kablooie!" Dougdoug blew up an imaginary planet.

"You know what, kid? These last three minutes, I'll never get them back."

"Where can we put it?"

"What?"

"This." The young man held up the spider.

"You're not reading the memo, Dougie. Let's go!" I showed Dougdoug the closest drain. Dougdoug scowled and walked over to the wall. He placed the insect behind the warm heat pipe.

I muttered. "It's a spider."

"It's not... just that spider."

"You know, kid, with that live-and-let-live attitude, I'm afraid you're not going too far in the oil industry. Besides, this dig-and-dump is the old technology. The new way is to leave everything on top, drill sideways, and suck the shit out. It's spider-approved."

Dougdoug looked at me, then laughed. He said, "We're digging ourselves a huge karma debt here. Payback's gonna be a bitch."

"Once you been here for a while, kiddo, you get to know who's lying to you, and it ain't always the oil companies. But for what it's worth, Miss Muffet, you've saved a spider."

We walked for a bit. Finally Dougdoug mumbled.

"The Egyptians have their pyramids, we got tar ponds."

"Kid?"

"Yeah?"

"It was a spider."

DAY TWENTY-FOUR

(The Great Eastern)

"Don't ya love that sound," Jason said.

"What sound?" Pops asked, looking up from Gwen's desk. "Oh, that." Pops stared at the wall in the direction of the coker, listening to the rattle of impact wrenches closing up man-ways. "When is Gwen getting back?"

"Haven't a clue." Jason said. "Once we got the call about Lobotomy—ah, Tim—was in the hospital, she grabbed her stuff and took off. Said something about running interference. Took the truck too."

"They were close?" Pops asked.

"Apparently."

The two men stopped to listen to the staccato metal-on-metal bang of another impact gun, tightening metal nuts the size of fists until the steel almost distorts, closing up the man-ways, getting the men ever closer to home. Jason smiled.

"You know, Jay," Pops said, "twenty-four days ago, if you walked inside the coker and saw the broken steel trays, walls with huge divots in them, metal in shards, you'd have thought, How the hell are we ever going to finish all this in twenty-four days? Then the guys start working like ants on a sandwich, and somehow it all comes together. She's done." Pops set down his coffee mug by the computer, creating another coffee ring on top of Gwen's spotless desk.

"You're lucky Gwen's away," Jason said, looking at the desk.

"A guy told me once that a bunch of years ago in Sarnia, he and his buddy were working together on Tower Fourteen. As he comes out of the tower for the last time, the mechanics were sealing up the man-ways. He turned to his friend and said, 'Where's the tools?' They look at each other, then back at the tower. They didn't tell anybody that they had left the tools inside the tower. That's not something you'd advertise. Even the guys would give you static, not to mention the bosses."

"Yeah, no shit," Jason said.

"So, a year passes. There's another shutdown in Sarnia. These two guys are working together again. One says, 'I wonder if those tools are still there?' So just after the crew opens the man-ways again, they climb back inside."

Jason stared at Pops.

"All that was left was the galvanized pail's bottom ring where they had put it down. What was left of the flashlight was just wires, and an outline on the steel where the batteries once were. And the hand wrenches? They were all shiny and clean, laying there, just like new."

"What did they do with the wrenches?" Jason asked.

Pops shrugged. "Put 'em in another pail."

The old welder leaned back and put his feet on Gwen's desk.

"I used to weld in the double bottoms of icebreakers in the Arctic. The ship was built with one hull inside another hull, so if a hole gets ripped in the outside hull, the ship won't sink. Well, that's the plan, anyway. The space in between the outer hull and the inside hull is divided up into small—I mean really small compartments. Each compartment was about the size of four coffins. You could lie down in them, but you couldn't stand up, not fully. The only thing you could do is kneel, kneel, and crawl.

"Everybody got a compartment to weld. One compartment, one lightbulb. Shit, it was dark. Dark, smoky, noisy, and real easy to get scared. I spent three months singing to myself so I wouldn't panic. I had this repertoire of songs I would go through. It got to the point that once I went through all the songs and got to 'Mary Had A Little Lamb,' I knew I was close to freaking out. Time to get out and check the welding machine, whether it needed it or not."

"'Mary Had a Little Lamb'?" said Jason.

"Yeah."

"That's not a song."

"I know my nursery rhymes."

The office went silent, except for the sound of two men humming.

Jason coughed and looked around. "Did you ever hear about the *Great Eastern*?" he asked.

"No."

"The *Great Eastern* was a steel-hulled ship that was made in England. Men were still wearing stovepipe hats. Back then they riveted the ship together, no welding. The shipyard rattled so much with all those riveting guns going that everybody was deaf by the time they were thirty.

"The *Great Eastern* was so underpowered it could barely get out of its own way. When they launched the ship, the wave came across the river and drowned a couple of people. The architect who designed the ship died young. The shipyard that made the ship went bankrupt. The company that owned it went under. For twenty years, everybody who even came close to the *Great Eastern* suffered. It was a jinxed ship.

"The only time the *Great Eastern* made any money was as a storage ship for the first telegraph cable laid under the North Atlantic. No other ship in the world had enough room to hold all that cable.

"After that trip, they sold it for scrap. When the guys in the scrapyard was tearing the *Great Eastern* apart, way down at the bottom they found the skeletons of a boilermaker and his apprentice. They had been accidentally sealed up into the hull when they first made it."

Jason's face was still. His eyes watched Pops.

"Can you imagine?" Jason said. "You're down there, in the black, and when you go back the way you came, now there's a wall. At first you can't believe it's happened. They can't hear you scream because they're making so much noise with the riveting. You hit the wall with your tools, your fists, anything. But the whole ship is vibrating with noise. You scream and scream. Nobody can hear you. You kneel by that wall, and die in the black. If you're lucky, the compartment is airtight and you die within a couple of hours."

"If not?" Pops said.

"Have you ever seen a guy who's suffocated?" Jason asked, grimacing.

For a long time, the only sound was the static crackle of the two-way radio. Then the distant rattle of an impact gun filled the trailer.

Jason jumped up so quickly, Pops started. "I'm going to check on the men."

"Yeah," Pops said. "Count 'em. Twice."

(Lonesome Road)

There are landscapes in the memories of my long-agos, points of reference in the graph line of my life. There was that day when the spray of a sunset caught the yellows of a Saskatchewan wheatfield. I stopped the truck just to watch God end the day. Once, after a shutter in McMurray, I was surprised by a herd of bison in a green field south of Cold Lake. The herd lay in the tall blue grass with their heads up like they were rising out of the flax field, going towards the matching grey green thunderheads above. There's that first sight of those broken-toothed Rockies along the horizon west of Calgary. It's the colours and the smells of the land that I remember, only the land.

I hear a song, and memories come flooding back. As the music plays, I would instantly be transported from freezing in this Fort McMurray winter, to driving into the orange sunset down a Saskatchewan gravel road in July, the smell of newly-cut hayfield warm in my nose, the sun on my face, and the prairie wind roaring in my ears, covering the rattle of the crickets, the birds, the truck's radio.

But there are few happy faces in my memories, only memories of the relief of leaving a tension-filled house, and the guilt of being an absent father.

The music would play, and I would get homesick, but I couldn't go home.

"Where you going from here?" the crew would ask.

"Ah, I'll try to grab another shutter in Fort Saskatchewan. I hear there might be a sixty-day shutdown in Regina. Maybe I'll even try to get on with that nuke plant in Ontario. I hear guys are making a hundred thou a year."

"Where's home?"

"Winnipeg."

I stay in Winnipeg on the chance I might get invited to one of my kids' homes, where I would see my grandchildren. The last time it happened, a wet-diapered cherub had climbed onto my lap and nestled there for a glorious half hour. I lived for a week with the memory of the smell of that angel's hair.

Jason spoke to the crew as they dressed. "Everybody comes back here at last coffee. If you're not here for last coffee, I'm not paying you for the day. Stay until last coffee, and I'll pay you for the entire shift."

"Layoff's payoff?" I asked.

"You know you only get a cheque at layoff time if the job's less than a week," Jason said.

"Worth a shot. Can we take a long coffee break?"

"Don't push it." Jason smiled.

"Roll back all the welding cables except one. Roll back all the air hoses except one. Bring all the tools down to the tool room. Leave enough tools to close off that last man-way."

"When's the engineer coming to do the final inspection?" Double Scotch asked.

"After lunch. I told him if he's late, he'll button up that last man-way himself."

The crew bundled up one last time. They shuffled towards the coker's stairs like a line of dirty blue penguins climbing a black ice floe.

I watched them go and thought,

These scraggly hunched men with their ripped and muddied coveralls are all that's left. Injuries, violence, and personal disasters so prevalent in construction it's almost trivial. We started with over twenty men three weeks ago, and we end up with what? Less than enough warm bodies to make up a good poker game. Typical shutdown.

Tonight they'll go home to their families and for the first week it'll be like Christmas, and a honeymoon, and winning a lottery all rolled into one. They'll have screaming monkey sex, dine out like corporate execs, try to rid the world of alcohol, all for a week.

Then one morning, he and his old lady will wake up, and something will set one of them off. They'll have a huge screaming match. The kids will cry, the dog will bark, and the guy will realize that he's only a disruption in their lives. They don't want him back. That's when he either starts looking for a steady-Eddie job in town, or he starts filling that duffel bag that never left the foot of his bed.

He'll pack his duffel bag and go down to the union hall for a job, any job. Going back to being that distant, perfect father rather than being that houseguest, hanging around bothering everybody.

He'll accept a job in a new boiler off towards the lakes, or an oil tank farm north of McMurray. Next morning, when his wife wakes up, the bed will smell of him, but it'll be empty. She'll get the kids off to school, wash the sheets, and get back to normal.

The day inched along. The engineer came and went. The cables, airhoses, and tools were packed in toolboxes. The men hung around the heaters on the seventh floor, waiting.

At ten minutes to three, the lunchroom door slammed open. With shouts and laughter, the crew exploded into the lunch trailer, throwing off coveralls, boots. Any gear that belonged to the company was flung into corners, the floor, anywhere.

Men frantically gathered their winter clothes, lunchboxes, personal items, and ran out the door with it all, like a TV game show where housewives try to gather as much money as they can carry in their chubby arms.

"See ya on the next one!"

"I was looking for a job when I found this one!"

"Hey! Give me a ride to town, wouldja?"

"I know the first thing you'll do when you see your old lady, but what's the second thing you'll do?"

"Put my suitcases down!" everyone shouted in unison.

"Whaooo!

"Rick! Goodbye!"

"See ya!"

"See ya, see ya on the next one!"

The last man ran out as the door slammed, shaking the trailer. The trailer, still vibrating from the shouts of the departed men, went silent. After a long time, the sounds of the refinery crept back.

The foreman's voice startled me out of my reverie.

"Rick, it's been good working with you."

"Thanks, Jay. You too."

"You can go if you want."

"I'm in no rush."

"When you going to slow down, enjoy your pension?"

I rose from the desk and gathered up my winter parka.

"Damned if I know, Jay. Damned if I know."

EPILOGUE

The hum from the tires got louder as that Fort McMurray radio station buried itself in snow and static. My eyes squinted into the mid-afternoon sun and the treed horizon, the long shadows a preview of the night.

Going south.

This truck is my home.

I wondered about the men I worked with. One by one, the faces will fade into the slow darkening of past jobs until those men will just be a vague feeling—at best a flicker of scenes, like some jerky silent movie.

If I meet one of my crew within a year, we'll laugh and slap each other on our backs and ask, "Whatever happened to...?"

If I see him after a couple of years, I might be able to remember his first name. I'll introduce myself and smile a weak smile. We'll talk awkwardly, and both of us will be glad when we have to go someplace, away.

Longer than that, the man's name will disappear into that fog of time and travel.

I let up on the gas pedal. I thought, "Why push it? Where do I have to be?"

I thought I might rent a room in that Alberta border town where

I stayed once. The house faced into the sun, dug into a bald hill with parallel lines sawed into it by the hooves of generations of cattle. The back room of that apartment had dark curtains and a wide-screen TV, and smelt faintly of bleach over mould. But it was cheap and warm, and the owner was friendly.

But really? This truck is home.

Journeys cleanse you. The longer and more arduous the trip, the more you are purified. The rumble of the engine's cylinders and the wind in the windows washed away the smells, sounds, and feelings of Fort McMurray. Except to fill the tank, I didn't need much of an excuse to just keep right on driving.

This truck is home.

When the Greek gods finished toying with that ancient Jason, they told him to pick up an oar from his ship, the *Argo*, and carry it far inland.

Once he had walked far enough away from the sea he would meet a man on the road who had never seen the ocean. When the passer-by asked him what manner of spear he carried, the gods instructed Jason to immediately plant that oar in the ground. From that oar, an olive tree would grow, the oil of which would sustain Jason for the rest of his life.

I'm finished with Fort McMurray, for now. I'm driving south. When I'm at a motel or a gas station and a passer-by casually asks me why I have an electrical cord coming out of my truck's grill, I'll stop.

Also by Rick Ranson

Working North: DEW Line to Drill Ship
978-1-896300-73-3, $19.95 CDN/$15.95 US

In the off hours of jobs in the Arctic, workers have a choice: to gamble, drink, or watch mind-numbing television. Ranson chose to write letters home describing the daily events from his experiences working on drill ships, construction sites, and DEW Line radar stations in the Arctic. *Working North* is a collection of these fascinating stories. The tales told by Ranson include staking out a polar bear, reviving a deserted ship, conflicts with racist coworkers, and welding in the bowels of a sinking ship.

Paddling South: Winnipeg to New Orleans by Canoe

978-1-897126-23-3, $19.95 CDN/$16.95 US

In the Fall of 1969, Rick Ranson and John Van Landeghem, both barely out of high school, took on the might of the Red and Mississippi Rivers to paddle a canoe from Winnipeg, Manitoba, to New Orleans, Louisiana. Combining high drama with hilarity, Ranson tells how the duo ducked bullets in St. Louis, avoided a whirlpool, worked on a Mississippi tow boat, sailed a yacht through a barge-congested Cairo, IL, and spent a few days in the Fargo City Jail, all while meeting an eclectic array of unforgettable characters. *Paddling South* tells the incredible tale of how they survived the three month trip on the often treacherous rivers, beset by snow storms, hurricanes, monstrous waves, and unseen dams.

"His story is incredibly intriguing and entertaining, filled with plenty of entertaining characters. A solid and hilarious story about two friends right out of high school, Paddling South is one of the most entertaining memoirs one could find."
~ MIDWEST BOOK REVIEW